P9-DXJ-142

Books are not men and
yet they are alive,
they are man's memory
and his aspiration.
Stephen Vincent Benet
Mark E. Miller

Politics and the Restraint of Science

LEONARD A. COLE

ROWMAN & ALLANHELD
Totowa, New Jersey

ROWMAN & ALLANHELD

Published in the United States of America in 1983
by Rowman & Allanheld, Publishers
(A division of Littlefield, Adams & Company)
81 Adams Drive, Totowa, New Jersey 07512

Library of Congress Cataloging in Publication Data

Cole, Leonard A., 1933–
Politics and the restraint of science.

Includes index.
1. Science—Social aspects—United States.
2. Science—Political aspects—United States.
3. Science and state—United States. I. Title.
Q175.52.U5C653 1983 500 83-2992
ISBN 0-86598-125-6

83 84 85/ 10 9 8 7 6 5 4 3 2
Printed in the United States of America

For Ruth

Contents

List of Tables

Preface

If, as Max Weber observed sixty-five years ago, everyone lives "for" politics or "off" politics, no less can now be said about science. As the effects of science on society continue to grow, the inextricable ties between political and scientific activities become increasingly evident. Many people are uncomfortable about this, for science is commonly regarded as a more lofty enterprise than politics and mixing the two, accordingly, could only be to the detriment of science. This book has grown out of my interest in this premise and in the arguments that challenge its validity.

In exploring these issues I have profited from discussions with more friends and colleagues than I can recount. In particular I thank Joseph Haberer, Loren Graham, Allan Mazur, Marlan Blissett, Brian Barry, and Stephen Shalom, who read all or part of the manuscript.

In the course of research, I spoke with more than one hundred scientists and others concerned with the making of science policy. Listing them would be too cumbersome, but I am grateful to all. Several were unusually helpful in sharpening my understanding of the nuclear and recombinant DNA issues, often by their advocacy of singular positions. I thank among them Paul Berg, Peter Bradford, Liebe Cavalieri, Erwin Chargaff, Roy Curtiss, Daniel Ford, Henry Kendall, David Lilienthal, David Okrent, John Palfrey, John Pastore, Robert Pollard, Melvin Price, Isadore Rabi, Fred Reines, Marcus Rowden, Robert Serber, Maxine Singer, Robert Sinsheimer, DeWitt Stetten, Bernard Talbot, George Wald, and Alvin Weinberg.

I acknowledge with gratitude permission to use portions of material that I wrote for *The Sciences* (December 1981) and *The Nation* (October 23, 1982). I am indebted most of all to my wife Ruth, whose wisdom and support have benefited this study from beginning to end.

PART ONE
Introduction

1

Restraining Science: Debate and Dilemmas

> What is science? Truly a big question. . . . I shall give a simple answer. Science is the attempt to learn the truth about those parts of nature that are explorable.
>
> *Erwin Chargaff*[1]

Since 1633, when Galileo was forced by the church to renounce his belief that the earth revolved around the sun, scientists have been wary of outside authorities seeking to impose scientific "truth." Suspicions were rekindled more recently when Stalin graced Lysenko's version of biological truth with the Communist Party's imprimatur. Genetics became outlawed, geneticists pariahs. When the Nazis murdered "inferior" humans in the name of racial science, the consequences of politically dictated scientific truth reached a horrifying nadir.

Memories of these historic cases continue to underlie suspicions that governmental interference with any scientific activity threatens to pervert all scientific activity. Yet during the past decade, some observers have become convinced that the character of science has changed. They worry that unregulated science and technology threaten the human population with irreversible hazards. Unlike any period before the nuclear age, scientific and technological activities have come to be seen as potential vehicles of disaster that can no longer be controlled by the will of humans. Thus, the dangers of radioactive waste, indestructible chemical toxins, and recombinant DNA research became issues of acute concern during the seventies.

Debates during the decade focused on how to reconcile the scientist's right to pursue his work with the government's responsibility to protect the citizenry. While many continued to argue against any effort to regulate science, others began to question this traditional position. By the end of the decade, several conferences had been convened in an effort to assess the role of political authority relative to science. Scientists, lawyers, social scientists, and philosophers debated at conferences whose titles emblazoned the themes of concern: "Regulation of Scientific Inquiry," "Limits of Scientific Inquiry," "Scientific Expertise and the Public," "Fear of Science—Trust in Science," and "Science and Ethical Responsibility."[2]

The conferences dealt with the full breadth of scientific activity—from inquiry to application, from discovering knowledge to utilizing it for the benefit of society. Some emphasized historical perspectives, others technological aspects. But the central question remained: Who should control scientific activity? Despite the increased attention to the issue, it appeared no nearer to resolution at the beginning of the eighties than before.

The conferences underscored an increasing recognition of the unavoidable tie between politics and science. But while politics and science have always influenced each other, concerns expressed at the conferences and elsewhere frequently revealed confusion and misinformation about the roles played by political systems. In expressing their anxiety about political interference with scientific activity, many observers equated restraints on scientific activities in the United States with the impositions prompted by the political systems in Nazi Germany, Stalin's Russia, and seventeenth century Italy.

In arguing against external controls, Gerard Piel, typically, raised the specter of the Galilean, Lysenko, and Nazi episodes. "In our country today," warned Piel after citing the three historic cases, "those [who] call for regulation and restraint of scientific enquiry must recognize they are tampering with the moral and pragmatic foundations of our freedom and welfare."[3] Mindful of the Inquisition's "constraining Galileo to abjure from inquiry," DeWitt Stetten echoed the same theme and argued that failure to include a right to freedom of inquiry in the United States Constitution "is probably only an historical accident."[4]

Others were more specific with their analogies. The imposition of Lysenkoism on Soviet society was repeatedly equated with efforts to regulate recombinant DNA research (gene-splicing).[5] One scientist drew an extravagant list of demons that he held comparable to governmental regulation of recombinant DNA experimentation, including the actions of "the Holy Inquisition, witch-hunting, the tribulations of Copernicus and Galileo Galilei and, more recently, the infamous Scopes trial and the Lysenko affair."[6]

Elsewhere, parallels have been drawn between Nazi racial science and a panoply of contemporary biological issues, including genetic intervention;[7] fertilization of human ova outside of the body;[8] restraints on, or imposition of medical therapy;[9] and a variety of questions regarding biomedical ethics.[10]

Similarly, the church's persecution of Galileo has been cited as analogous to a variety of incidents in which American scientists were victimized by governmental agencies. The ghost of Galileo was summoned to describe the stifling of scientists who argued that existing standards of radiation were unsafe;[11] who objected to proposed weapons development;[12] and who urged restraint of recombinant DNA research.[13] "Twentieth-century Washington," observed Weisband and Franck, has become comparable to "the Vatican of Galileo's time."[14]

Some critics undoubtedly engaged in hyperbole for dramatic effect. Others consider the analogies more than mere caricature. Most of 632 scientists who were surveyed in conjunction with this study take the matter seriously. The format of the survey is explained in the Appendix, and discussion of this issue is amplified in later chapters. But we note at this point that 57 percent of the respondents believe that the American political system is quite capable of perverting science in a manner comparable to the three historic cases.

Such perceptions raise worrisome questions. If the perceptions are accurate, every action taken under political auspices that relates to science must be a cause of suspicion. If the American political system is potentially as wicked as those in the historic cases, presumably the only safeguards against equivalent political impositions on science must lie with scientists. Since scientists are the discoverers of scientific truth, according to this line of reasoning, only they would be in a position to protect us from truth imposed by political authority. This follows from assumptions like Bronowski's that scientists "rest their scientific faith on an uncompromising adherence to the truth, and the irresistible urge to discover it. All of them spurn that grey appeal to expediency. . . ."[15]

The principal argument in this study holds that these perceptions are not valid, that the American political system, as presently constituted, protects both scientists and the public from perversions committed in the name of science as occurred in the historic cases. Neither historical analysis nor the responses of contemporary American scientists in interviews and surveys support the notion that scientists would act as protectors against state-imposed scientific truth.

The implications of this with respect to the proper role of the political system is central to the current debate. Rather than fear governmental interference, scientists and the public should welcome informed governmental restraint of certain scientific activities. In developing this theme, this study will delineate features within the political systems of the historic cases that enhanced the likelihood of politically imposed scientific truths and then make comparisons with American political institutions. Only after identifying such features may one feel more secure or vulnerable about the governance of science, for political systems, depending on their character, may protect or abuse their citizens in matters of science as in other affairs.

Additionally, different aspects of scientific activity will be discussed, while recognizing that some threaten the welfare of the citizenry and some do not. The supposition that every political system carries a potential equal to every other for wicked impositions is naive, despite the beliefs of some observers to the contrary. No less naive is the claim that all scientific activity is harmless and that none requires external restraint.

Restraint

Restraint of science by political authority may be relative or absolute. Only absolute restraints will be of central concern in this study. Relative restraints involve inhibition, but not prohibition, of certain scientific activities. As governmental agencies sponsor research and development, they may promote one avenue of inquiry at the expense of another. Some scientific projects may be deprived when funds are diverted for military research or for development of new sources of energy. Government-sponsored prizes and appointments induce some scientists to avoid activities that do not draw funds and recognition. Such activities are not forbidden, and a scientist retains choice in his professional pursuits. We are concerned here, however, with absolute restraints, the principal subjects of the debates through the seventies and into the eighties.

Absolute restraints entail total prohibition by political authority of some scientific activities. Three areas were often cited, though not always delineated, during the debates. Failure to differentiate among the three contributed to confusion about the legitimacy of any restraint. The first absolute restraint involves the imposition by political authority of scientific truth or fact, and the prohibition of alternative considerations. The three historic cases stand as models.

The second absolute restraint concerns the prohibition of scientific inquiry because of ethical considerations. Disclosures during the 1970s that governmental agencies had sponsored dangerous experiments on unwitting human subjects lent particular poignance to this issue. The third arises from the prohibition of certain scientific or technological activities because of their risk to public safety. The perceived dangers associated with nuclear power, the use of indestructible chemicals, recombinant DNA research—and what to do about them—fall into this category.

This book will not only examine the three types of absolute restraints, but assess the threat of certain activities, and the consequences of their restraint, to the welfare of American society.

The Lines of Inquiry

In developing the contention that analogies between incidents in American society and the three historic cases are inappropriate, the next few chapters will examine the character of science and the political systems in each society. In Chapter 2 the Galilean, Lysenko, and Nazi cases of imposed scientific truth are reviewed. Chapter 3 describes the role of scientists in each incident. It recounts how most scientists supported or acquiesced in the perversion of their science.

Chapter 4 examines a range of explanations for the compliance of most scientists with politically imposed scientific truth in the historic cases. The conclusion is drawn that scientists, like other citizens, are possessed of an inner requirement to obey authority—even when the authority mandates "truths" known to be false. Chapter 5 reviews science and the political systems in the societies of the historic cases and compares them with institutions in the United States. The argument is further developed that the American political system precludes the kind of impositions that occurred in the other three societies.

The remaining chapters concentrate on contemporary American science and politics. While rejecting the possibility in American society of imposed scientific truth or fact as in the historic cases, the discussion will focus on the other issues involving absolute restraint: ethical contraventions and potentially catastrophic scientific and technological activity. Chapter 6 discusses responses of scientists to a series of issues and inquires into their views about the nature of scientific truth and its perversion, and about the ability of scientists to analyze issues compared to nonscientists. The chapter then examines nascent efforts to impose scientific truth within American society. The antievolution campaigns of the fundamentalists and creationists are reviewed, as are recent efforts to legislate when human life begins. The argument is reinforced that the overall political system, not scientists, ultimately stands as protector of the citizenry.

Chapter 7 reviews incidents of life-threatening research on unwitting humans by the Public Health Service, the Central Intelligence Agency, and the Army. It demonstrates the need for restraining this kind of scientific inquiry, despite cries by some about a scientist's right to freedom of inquiry. Protection offered by political institutions is reviewed, and the need for additional safeguards is emphasized.

Chapter 8 reviews the politics of nuclear power and recombinant DNA research. Regardless of technical arguments about the benefits or dangers of either enterprise, the political system responded to each very differently. The history of nuclear policy represents a tale of distortion of traditional political institutions and has accounted, in part, for the continuing ferment about the issue. In contrast, most policy questions about recombinant DNA were resolved because traditional political institutions were permitted to function. The chapter is structured largely on information obtained from interviews with scientists who have been involved with nuclear and recombinant DNA affairs.

Chapter 9 turns to the question of the legitimacy of governmental interference with scientific activity and is organized about the responses of scientists to this study's survey. Their answers and comments reveal both a sense of need and apprehension about governmentally imposed restraint of any scientific activity. In the perspective of these responses, a proposal is offered for an institutional arrangement that should help address the problem. A new Science Hearings Panel could enhance the functioning of the political process toward protecting the citizenry from potentially disastrous activities.

The concluding chapter cites the increased skepticism by younger scientists about a range of contemporary questions. Younger scientists, according to the survey, are less tied to traditional beliefs about the nature of science and are more likely to accept governmental interference with potentially hazardous scientific activity. The implications of this finding for the future of American science are discussed in the perspective of the entire study.

Determining the appropriate relationship between science and external authority is of paramount contemporary concern. Yet the issue has been long-standing. It was central three centuries ago as well, during the confrontation between Galileo and the church.

Notes

1. Erwin Chargaff, *Heraclitean Fire, Sketches from a Life before Nature* (New York: The Rockefeller University Press, 1978), p. 156.

2. Keith M. Wulff, ed., *Regulation of Scientific Inquiry* (Boulder, Col.: Westview Press, for the American Association for the Advancement of Science, 1979); "Limits of Scientific Inquiry," in *Daedalus,* vol. 107, no. 2 (Spring 1978); Hans Skoie, ed., *Scientific Expertise and the Public,* Conference Proceedings

(Oslo, Norway: Institute for Studies in Research and Higher Education, The Norwegian Research Council for Science and the Humanities, 1979); Andrei S. Markovits and Karl W. Deutsch, eds., *Fear of Science—Trust in Science* (Cambridge, Mass.: O. G. and H. Publishers Inc., 1980); Sanford A. Lakoff, ed., *Science and Ethical Responsibility*, Proceedings of the U.S. Student Pugwash Conference, University of California, San Diego, June 19–26, 1979 (Reading, Mass.: Addison-Wesley Publishing Co., 1980).

3. Gerard Piel, "Scientific Research: Determining the Limits," in Wulff, pp. 43–45.

4. DeWitt Stetten, Jr., from transcript of forum on *Experiments and Research with Humans: Values in Conflict* (Washington, D.C.: National Academy of Sciences, 1975), p. 200.

5. John Lear, *Recombinant DNA, The Untold Story* (New York: Crown Publishers, 1978), p. 157; George H. Kieffer, *Bioethics: A Textbook of Issues* (Reading, Mass.: Addison-Wesley Publishing Co., 1979), p. 423; *The New York Times*, July 24, 1977, p. 18-E.

6. Waclaw Szybalski, "Dangers of Regulating the Recombinant DNA Technique," *Trends in Biochemical Sciences*, vol. 3, no. 11 (November 1978), p. N 243. Similar remarks were made at a conference sponsored by the American Association for the Advancement of Science, titled "Law/Science Perspectives on Public Policy Making," January 25–27, 1979, Airlie, Virginia, in conversations with Szybalski.

7. Amitai Etzioni, *Genetic Fix, The Next Technological Revolution* (New York: Harper Colophon Books, 1975), pp. 110–11.

8. Walter Sullivan in *The New York Times*, December 12, 1978, p. C-2.

9. Thomas Szasz, *The Theology of Medicine* (Baton Rouge, La.: Louisiana State University Press, 1977), p. 102.

10. "Biomedical Ethics and the Shadow of Nazism," Special Supplement in *The Hastings Center Report*, vol. 6, no. 4 (August 1976), pp. 1–18.

11. Richard S. Lewis, *The Nuclear-Power Rebellion* (New York: The Viking Press, 1972), p. 49.

12. Joseph Haberer, *Politics and the Community of Science* (New York: Van Nostrand Reinhold Co., 1969), pp. 250–51; Giorgio de Santillana, *The Crime of Galileo* (Chicago: University of Chicago Press, 1955), p. viii.

13. Nicholas Wade, *The Ultimate Experiment, Man-Made Evolution* (New York: Walker and Co., 1977), chap. 9.

14. Edward Weisband and Thomas M. Franck, *Resignation in Protest* (New York: Grossman Publishers, 1975), p. 121.

15. J. Bronowski, *The Common Sense of Science* (New York: Vintage Books, n.d.), p. 123.

PART TWO

The Restraint of Science, Historical Perspectives

2
The Historic Cases

The Galilean Affair

Until the sixteenth and seventeenth centuries, with the exception of a few ancient Greeks, people believed that all celestial bodies rotated about a stationary earth. For Christians the assumption that the earth was the center of the universe was based not only on common sense, but on church doctrine. To claim otherwise was to challenge a tenet of faith, the word of God.

Catholic philosophers and scientists before the sixteenth century relied on the writings of Aristotle and the second-century formulations of Ptolemy to explain the heavens. The sun, the moon, the planets, and stars revolved about the earth, according to Aristotle. Their orbits were perfectly circular, and the essential nature of celestial material was different from terrestrial.

Ptolemy elaborated on Aristotle's model, and to explain the most glaring contradiction—the apparently noncircular paths of the planets—he postulated an ingenious series of epicycles. A planet would make tiny circular orbits while traversing its larger orbit encircling the earth. These explanations provided the foundation for astronomy for fourteen centuries. In 1543, with his publication of *Revolutions of the Celestial Orb,* Nicoleus Copernicus revised the Ptolemaic system. By suggesting that the sun was at the center of the "universe," and that the earth was one of its orbiting planets, as well as rotating on itself every twenty-four hours, Copernicus simplified the complicated system of epicycles.

Few observers were convinced of Copernicus's proposals. His treatise was mathematically complex and was commonly perceived as a clever technical exercise, not as a valid description of reality. Copernicus's revised universe appeared to threaten neither established scientific understanding nor church dogma. So the matter stood for more than half a century until Galileo Galilei joined the issue.

GALILEO'S CHALLENGE

Galileo was born in Pisa in 1564, twenty-one-years after Copernicus published his treatise. By his early twenties he had made several discoveries in mechanics, and in 1604 he devised his classic theory on the motion of falling bodies. He held lectureships at the University of Pisa and later at the University of Padua, but he was little known beyond local circles. In March 1610, Galileo was catapulted into international prominence with his publication of *The Starry Messenger,* a small book pronouncing his inventive use of a "spy glass" to observe the heavens. Galileo's text sparkles with excitement about the "universe that I enlarged a hundred and a thousand times from what the wise men of all past ages had thought."[1]

Galileo's telescope placed before the eye irrefutable evidence that old beliefs were untenable: Mountains and valleys on the moon, seen for the first time, challenged traditionally held distinctions between celestial and terrestrial terrains; the phases of Venus, visible only with the aid of the telescope, sustained the Copernican supposition that Venus had to exhibit phases like the moon's because of its orbital location between the earth and the sun; Jupiter's four newly discovered moons could be seen revolving about the larger planet as if they comprised a small-scale solar system; observation of sun spots confirmed the sun's rotation on its axis, thereby contradicting established canons of solar perfection and immutability. "This novelty," conjectured Galileo about his telescope, "may well be the funeral, or rather the last judgment, of pseudo-philosophy."[2]

Galileo was soon disabused. While his invention excited some observers and drew support for his Copernican ideas, many appeared to reject what their eyes beheld. Father Clavius, a Jesuit astronomer, could not believe that there were mountains on the moon. Though he later relented, at first he insisted, as did others, that the apparent irregularities were illusory.[3] The leading philosopher at Pisa, Giulio Libri, believed that since Aristotle had said nothing about Jupiter having moons, the moons were an illusion.[4] Father Scheiner, a Jesuit mathematician who embraced the Aristotelian dogma of solar immaculacy, argued that since the sun could not rotate, the sunspots had to be small stars circling the sun.[5] Several scholars refused even to look through the telescope.

A chastened Galileo wrote to his friend and fellow-Copernican of his frustration: "My dear Kepler, . . . what would you say of the learned here, who, replete with the pertinacity of the asp, have steadfastly refused to cast a glance through the telescope? What shall we make of all this? Shall we laugh, or shall we cry?"[6]

During the next few years Galileo was torn between advice to abandon the controversy and desire to press on with his cosmological arguments. He continued to write of his heliocentric convictions in letters to his friends.[7] Meanwhile, the Holy Office in Rome had received reports of Galileo's letters, as well as apocryphal descriptions of Galileo's position. As the church became interested in the issue, Galileo set off for Rome in December 1615 to present his views accurately and, he hoped, convincingly. In Rome Galileo was warmly received in official quarters. His arguments, however, were not. The mood set by Pope Paul V, a doctrinaire and inflexible man, was unreceptive to a departure from traditional views. Cardinal Bellarmine informed Galileo that Copernicanism would remain scripturally unacceptable.

In 1623, after the election of Maffio Barbarini to the papacy, Galileo's hopes were rekindled. The new pope, Urban VIII, had been a scholar and an admirer of Galileo's. Encouraged by friends and supporters, Galileo became enthusiastic about the possibility that the church's attitude toward heliocentrism would become more flexible. He met with Urban and felt sufficiently encouraged to begin a major work that would validate Copernicanism. In 1624, at the age of sixty, he started to write his *Dialogue Concerning the Two Chief World Systems.* He worked on it sporadically for six years, and it was published in February 1632.

Realizing that Urban VIII and other church leaders would not tolerate an outright refutation of the established position, Galileo sought to make his case obliquely. Because the *Dialogue* avoids presenting the Copernican position as absolute, Galileo assumed that the church would not be antagonized. As long as the Copernican model might be allowed as a legitimate (if hypothetical) alternative, Galileo believed that its compelling logic would displace traditional doctrine. The mutual regard that he had shared with the pope further convinced him that his presentation would never be judged heretical. The initial enthusiasm about his *Dialogue* by the literary public undoubtedly sustained his confidence.

Galileo must have been staggered when, in August 1632, two months after the book's publication, its sale was ordered suspended by the Inquisition. On October 1, he was told to appear in Rome at the Holy Office. He delayed departure, pleading illness and advanced age; but he was advised officially by the congregation of the Inquisition in January 1633 that, unless he obeyed forthwith, he would be conducted "to the prisons of this supreme tribunal in chains."[8] On February 13, after twenty-five days of difficult winter travel, Galileo arrived in Rome.

THE TRIAL AND RECANTATION

Galileo's trial by the Inquisition has been recounted by many scholars.[9] It is high drama that has left indelible marks on the history of science, on the image of the Catholic church, and on the ageless story of confrontation between the individual and authority.

Although Galileo remained at the mercy of the Inquisition during the trial, at first he denied any wrong doing. Unlike Galileo, however, the Inquisition understood what the outcome of the trial would be before it started. Its members employed contrived evidence that supported their accusation that Galileo had been informed not to defend Copernicanism in any way whatsoever. They played on the memory of an aged (now seventy), infirm, and frightened man.

At his final hearing on June 21, 1633, according to the rules, the accused was questioned under threat of torture as to his real belief. Galileo replied: "I do not hold and have not held this opinion of Copernicus since the command was intimated to me that I must abandon it; for the rest, I am here in your hands—do with me what you please."[10] The next day the sentence by the Inquisition was read, prefaced by a long reiteration of the church's charge against Galileo. He would be forced to renounce his apparent heresy and would be condemned to the prison of the Holy Office "during our pleasure." Clad in the white shirt of penitence and forced to kneel before his assembled judges in a large hall in the convent of Santa Maria Supra Minerva, Galileo read the abjuration prepared for him:

> I, Galileo, son of the late Vincenzo Galilei, Florentine, aged seventy years, arraigned personally before this tribunal and kneeling before you, Most Eminent and Reverend Lord Cardinals Inquisitors-General against heretical pravity throughout the entire Christian commonwealth, having before my eyes and touching with my hands the Holy Gospels, swear that I have always believed, do believe, and by God's help will in the future believe all that is held, preached, and taught by the Holy Catholic and Apostolic Church. But, whereas—after an injunction had been judicially intimated to me by this Holy Office to the effect that I must altogether abandon the false opinion that the Sun is the center of the world and immovable and that the Earth is not the center of the world and moves and that I must not hold, defend, or teach in any way whatsoever, verbally or in writing, the said false doctrine, and after it had been notified to me that the said doctrine was contrary to Holy Scripture—I wrote and printed a book in which I discuss this new doctrine already condemned and adduce arguments of great cogency in its favor without presenting any solution of these, I have been pronounced by the Holy Office to be vehemently suspected of heresy, that is to say, of having held and believed that the Sun is the center of the world and immovable and that the Earth is not the center and moves:
>
> Therefore, desiring to remove from the minds of your Eminences, and of all faithful Christians, this vehement suspicion justly conceived against me, with sincere heart and unfeigned faith I abjure, curse, and detest the aforesaid errors and heresies and generally every other error, heresy, and sect whatsoever contrary to the Holy Church, and I swear that in future I will never again say or assert, verbally or in writing, anything that might furnish occasion for a similar suspicion regarding me; but, should I know any heretic or person suspected of heresy, I will denounce him

to this Holy Office or to the Inquisitor or Ordinary of the place where I may be. Further, I swear and promise to fulfill and observe in their integrity all penances that have been, or that shall be, imposed upon me by this Holy Office. And, in the event of my contravening (which God forbid!) any of these my promises and oaths, I submit myself to all the pains and penalties imposed and promulgated in the sacred canons and other constitutions, general and particular, against such delinquents. So help me God and these His Holy Gospels, which I touch with my hands.

I, the said Galileo Galilei, have abjured, sworn, promised and bound myself as above; and in witness of the truth thereof I have with my own hand subscribed the present document of my abjuration and recited it word for word at Rome, in the convent of the Minerva, this twenty-second day of June, 1633.

I, Galileo Galilei, have abjured as above with my own hand.[11]

Galileo spent the remaining eight years of his life under house arrest in his villa near Florence. He was forbidden to receive visitors except by formal authorization. Despite his great misfortune, which was compounded by the death of his daughter, by failing vision, and the infirmities of old age, Galileo produced several scientific pieces, including his classic on *Two New Sciences.* He died on January 8, 1642.

With Galileo's passing went the last scientific and cultural giant that had symbolized Italy's preeminence in the intellectual world. By the end of the seventeenth century, England had assumed that position, which it later shared with France and Germany, but never again did Italy.[12] Galileo's conviction by the Holy Office meant, in Santillana's words,

> the end of the whole scientific movement in Italy and, worse, of Florence herself. . . . [S]he had lost her intellectual life at the hands of the Holy Office. We can date back to that day, indeed, the time when Florentine civilization, which had carried the world since the thirteenth century, practically vanished from history.[13]

In the Galileo affair the church demonstrated absolute political power. It pressed on Galileo the full weight of a political system, without compromise, without recourse to appeal. Through its treatment of Galileo, the church imposed a scientific dogma on Catholic society that endured for centuries. Political and social institutions were unavailable to counteract the church's imposition. There was no forum in which opposing views could be discussed, let alone transformed into policy.

Is there sense in suggesting then, as many have, that incidents in American society have been comparable to the Galileo case? When Piel or Szybalski argue that regulating any scientific activity in the United States is equivalent to the persecution of Galileo, they offer a shallow understanding of the differences between the two political systems.[14] When Richard Lewis describes governmental pressures on skeptics about nuclear radiation policies as "analogous to the reaction of the Holy Inquisition to the assertions of Galileo Galilei that the earth revolves about the sun," he obfuscates more than he clarifies.[15]

When Santillana calls the Oppenheimer case an "exact analogy in structure, in symptoms and behavior" to the Galileo case, he overlooks important differences not only in the cases but in the two societies that harbored them.[16] As will be discussed, it was these differences that ultimately permitted vindication of Oppenheimer in this society, but not of Galileo in his. There are similar confusions in analogies made by scholars between Lysenkoism and American science.

Lysenkoism

Although Galileo was victimized by politically imposed scientific truth, three hundred years later in the Soviet Union another scientist, Trofim Denisovich Lysenko, assumed the role of victimizer. Lysenko first gained national attention when a *Pravda* article reported in 1927 that he had "solved the problem of fertilizing the fields without fertilizers and minerals."[17] While working at an obscure experimental station in Azerbaidjan, Lysenko claimed to have grown a crop of peas during winter that kept the fields green for cattle feeding. He was visited by a journalist from *Pravda*, and the article materialized from his interview.

About a year later Lysenko introduced his discovery of "vernalization." At a conference held in January 1929, he claimed to have developed a means to grow winter wheat in the spring and summer. Ordinarily winter wheat will not ear when sown in the spring, but Lysenko soaked winter seed in water and subjected it to prolonged chilling, the process he called vernalization. In consequence, according to Lysenko, winter grain planted in the spring yielded a better harvest than spring grain. A commission appointed by the Ukrainian Commissariat of Agriculture visited Lysenko and came away enthusiastic about his project. The Ukrainian Commissariat then ordered further tests of vernalization. But even before the tests had begun, the Commissariat had announced to the press that a solution to the wheat problem had been found.[18]

Lysenko's report on vernalization in January 1929 was generally ignored by agricultural specialists. Most, like N. A. Maksimov, regarded his work as offering nothing original.[19] But when the Ukrainian Commissariat began to show interest in Lysenko, agricultural specialists became attentive as well. In October 1929 the Commissariat transferred Lysenko to the most important center for agricultural research in the Ukraine, the All-Union Institute of Plant Breeding in Odessa. The move was accompanied by the Commissariat's announced enthusiasm for vernalization.

Journalists sought the views of specialists in plant physiology and wheat production. The specialists responded cautiously. They warned against extravagant expectations based on Lysenko's sparse proof. Yet, as Joravsky reports, "they framed these cautions and criticisms with inconsistent expressions of respect for the practicality of Lysenko and the agricultural bosses who had ordered large-scale production trials of vernalization."[20] Thus began a peculiar contest between science and pseudoscience that was not resolved for a generation. In promoting his new technics Lysenko sought to make vernalization and practical achievement the organizing principles of plant physiology. He linked his approach to that of I. V. Michurin, an aged Russian plant breeder. Skeptical scientists were challenged to show practical success or give up in favor of Lysenko's "agrobiology." Emboldened by support from the government, Lysenkoites began to demand the total abolition of classical genetics.[21]

Lysenkoites and scientists confronted each other at a conference called to air the issues in December 1936. The absurdity of Lysenko's position was exemplified by his insistence at the conference that he never "denied the existence of genes," while moments later he could "deny corpuscles, molecules of some special 'substance of heredity.' " In characteristic demagoguery he stated further, "we not only recognize but, in our view, incomparably better than you geneticists, we understand the hereditary nature, the hereditary basis of plant forms."[22] Some scientists rejected Lysenko's ideas about heredity as ignorant and confused. A. S. Serebrovski told the conference that Lysenkoism was "an attempt to throw us backward a century." N. P. Dubinin feared that if Lysenkoism prevailed, "modern genetics will be completely destroyed."[23]

At another conference in October 1939 called ostensibly to achieve reconciliation between Lysenkoism and science, but again unsuccessful, Lysenko continued his demagoguery: "In order to obtain a certain result, you must want to obtain precisely that result; if you want to obtain a certain result, you will obtain it. . . . I need only such people as will obtain the results that I need."[24] To this an anguished geneticist, Iu. Ia. Kerkis, could only reply, "I cannot understand that. It just doesn't fit in my head."[25]

The confrontation between Lysenkoism and science simmered throughout the years of Stalinist terror and into World War II. The pressures of war left the Lysenko-science debate in limbo. Ravages wrought by the war on the Soviet Union helped to blur the impracticality of Lysenko's techniques. Manpower was displaced, agricultural areas became battlefields, and national attention was focused on fighting the Germans. But Lysenko continued to raise the stakes of credulity. At first it was vernalization, then agrobiology, then rejection of genetics, ultimately dismissal of the theory of natural selection and reversion to Lamarckism. Lamarck's notion that acquired characteristics could be passed from one generation to the next was taken seriously by few scientists since the discovery of genetic laws in the nineteenth century. Environment, according to Lamarck, and now Lysenko, determined the hereditary character of life, not genes. (Lamarck, who died in 1829, did not of course know about genes.) Lysenko thus stood as a measure of how much absurdity the official leadership, the informed observer, the scientific expert would accept, even applaud.

Stalin's drive against cosmopolitanism in the natural sciences began in 1948. The drive lent decisive impetus to Lysenko's efforts and led to his gaining absolute control of all biological sciences. In July 1948 the Council of Ministers decreed that thirty-five new academicians, most of them Lysenkoites, were to be seated in the Lenin Academy of Agricultural Sciences. The following month Lysenko opened a week-long session of the Academy with a report reiterating his views and chastising classical genetics.

The session signaled the end of all opposition to Lysenko, for at the close, Lysenko announced that the party's Central Committee had endorsed his views. The Central Committee's endorsement meant that Lysenkoism had become party gospel. Lysenko's power in biological science became absolute.

Can anything similar happen in the United States? Some have argued that governmental interference with scientific activity has already brought American society to the brink of Lysenkoism.[26] As in analogies between American scientific activity and the Galilean case, such linkage appears farfetched and misleading. Yet the specter of Lysenkoism became almost epidemic among opponents of restrictions on gene-splicing research. "Laboratories across the country," wrote John Lear, despaired of "another Lysenko affair."[27] Kiefer noted that the "vision of Lysenkoism [has been] frequently invoked" among opponents of regulation, and that this has tended to obfuscate the issues.[28]

An understanding of what occurred in the Soviet Union during the Lysenko period and the current treatment of science in the United States makes such analogies appear silly, yet many remain convinced of their validity. When the political systems are analyzed comparatively in Chapter 5, the weakness of such analogies becomes even more evident. The third celebrated historic case, science under the Nazis, bears similar themes.

Nazi Science

Hitler's appointment as chancellor of Germany in 1933 signaled the beginning of government approval of racial science. To be sure, anti-Semitism influenced scientific questions before Hitler achieved power. The fact that Einstein was Jewish was sufficient reason for some to question his theory of relativity.[29]

But Hitler's accession to power officially legitimized a confusion of race with science and scholarship for the entire nation.

Leading figures in every discipline took the cue. "The fight against the Jews," wrote one scholar, "was no more confined to shabby tracts by unknown authors; it had made its entrance into the respectable world of Germany."[30] Erwin Bauer, a biologist of international reputation and director of the Kaiser Wilhelm Institute for Breeding Research, wrote in 1934 of his support for a eugenics program. At the time of his writing, the law pertained only to mentally defective persons. He presciently noted that this was merely a beginning. Other "inferior" people were soon to be included.

> Every farmer knows that should he slaughter the best specimens of his domestic animals without letting them procreate and should instead continue breeding inferior individuals, his breeds would degenerate hopelessly. This mistake, which no farmer would commit with his animals and cultivated plants, we permit to go on in our midst to a large extent. As a recompense for our humaneness of today, we must see to it that these inferior people do not procreate. A simple operation to be executed in a few minutes makes this possible without further delay. . . . No one approves of the new sterilization laws more than I do, but I must repeat over and over that they *constitute only a beginning.*[31]

Dr. Ernst Rüdin, for many years a professor of psychiatry at the University of Munich, Arthur Gütt, M.D., a high official in the Ministry of Interior, and Falk Ruttke, doctor of jurisprudence, co-authored the early Nuremberg laws on "race hygiene." They drew their arguments not from political or ideological bases, but, instead, sought to position the 1933 Law to Prevent Hereditary Illness in Future Generations in a scientific continuum originating with the works of Darwin, Mendel, and others. Purifying the race was merely applying to man "the natural laws discovered for plants and animals." Their applicability to man, according to the authors, had been "fully and completely . . . confirmed during the last three decades both through family research and through the study of bastards and twins."[32]

In logical sequence, Professor Martin Staemmler, another distinguished scholar, suggested a year later:

> Extinction and selection are the two poles around which the whole race cultivation rotates, the two methods with which it has to work. . . . Extinction is the biological destruction of the hereditary inferior through sterilization, the quantitative repression of the unhealthy and the undesirable. . . . [The] task consists of safeguarding the people from an overgrowth of weeds.[33]

In view of the distinctive philosophical and behavioral underpinnings of the United States compared to Nazi Germany, it hardly seems appropriate to compare Nazi racial science to scientific activity in the United States. Yet, as in the Galileo and Lysenko cases, such allusions are common. Sometimes glibly, sometimes with deep conviction, critics of one or another American policy argue that scientific activity in the United States is parallel to that of Nazi science.

Amitai Etzioni cites the frequent association made between any consideration of genetic intervention in the United States with the German effort to breed a master race.[34] Recently a major center that is devoted to the study of ethics, featured a program whose title related biomedical activities in the United States to biomedical activities in Nazi Germany.[35] Others have warned that restraint of scientific inquiry in America could lead to the "bestiality" that characterized the Nazi medical experiments.[36]

After reviewing the three historic cases, references by observers who compare them to activities in the United States, seem misplaced. In Chapter 5 the argument that such comparisons overlook the fundamental differences in the nature of the political systems will be developed.

Those who liken American scientific activities to those in Galileo's Italy, Lysenko's Russia, or Nazi Germany reveal a lack of confidence in the protective ability of the American political system. They imply that American political institutions offer little insulation against the perversion of scientific truth. The principal line of protection, according to this view, must lie with the scientist. In the next chapter the validity of this formulation is questioned, and in examining the response of scientists to politically imposed truth in each of the historic cases, little is found to encourage confidence that scientists will act as protectors.

Notes

1. Quoted in Giorgio de Santillana, *The Crime of Galileo* (Chicago: University of Chicago Press, 1955), p. 5. See Galileo Galilei, "The Starry Messenger," in *Discoveries and Opinions of Galileo,* translated by Stillman Drake (Garden City, N.Y.: Doubleday Anchor, 1957), pp. 21–58.
2. Quoted in Santillana, p. 26.
3. *Ibid.*, p. 22.
4. James Brodrick, *Galileo: The Man, His Work, His Misfortunes* (London: Geoffrey Chapman, 1964), p. 51.
5. *Ibid.*, p. 67.
6. Quoted in Santillana, p. 9.
7. Ludovico Geymonat, *Galileo Galilei: A Biography and Inquiry into His Philosophy of Science* (New York: McGraw-Hill Book Co., 1965), p. 58.
8. Geymonat, p. 141.
9. *Ibid.*, chap. 8; Santillana, chap. 12; Karl von Gebler, *Galileo Galilei and the Roman Curia,* translated by Mrs. George Sturge (Merrick, N.Y.: Richwood Publishing Co., 1977), chaps. 7–9.
10. Quoted in Santillana, p. 303.
11. Quoted in *ibid.*, pp. 312–13.
12. J. H. Randall, "The Newtonian World Machine," in Frederick Gentles and Melvin Steinfeld, eds., *Hangups from Way Back: Historical Myths and Canons,* Vol. 1, 2nd ed. (San Francisco: Canfield Press, 1974), p. 473.
13. Santillana, p. 305.
14. Gerard Piel, "Scientific Research: Determining the Limits," in Keith M. Wulff, ed., *Regulation of Scientific Inquiry* (Boulder, Col.: Westview Press, for the American Association for the Advancement of Science, 1979), pp. 43–45; Waclaw Szybalski, "Dangers of Regulating the Recombinant DNA Technique," *Trends in Biochemical Sciences,* vol. 3, no. 11 (November 1978), p. N 243.
15. Richard S. Lewis, *The Nuclear-Power Rebellion* (New York: The Viking Press, 1972), p. 49.
16. Santillana, p. viii.
17. David Joravsky, *The Lysenko Affair* (Cambridge, Mass.: Harvard University Press, 1970), p. 58.
18. *Ibid.*, p. 60.
19. Zhores A. Medvedev, *The Rise and Fall of T. D. Lysenko* (New York: Columbia University Press, 1969), pp. 13–14.
20. Joravsky, p. 62.
21. *Ibid.*, p. 97. On Lysenko's linkage to Michurin see Dominique Lecourt, *Proletarian Science, The Case of Lysenko* (London: NLB, 1977), pp. 43–46.
22. Joravsky, p. 104.
23. Loren R. Graham, *Science and Philosophy in the Soviet Union* (New York: Vintage Books, 1974), p. 216.
24. Joravsky, p. 110.
25. *Ibid.*
26. Piel, pp. 43–45; Szybalski, p. N 243.
27. John Lear, *Recombinant DNA, The Untold Story* (New York: Crown Publishers, Inc., 1978), p. 157.

28. George H. Kieffer, *Bioethics* (Redding, Mass.: Addison-Wesley Publishing Co., 1979), p. 423.
29. Ronald W. Clark, *Einstein, The Life and Times* (New York: World Publishing, 1971), pp. 256–58.
30. Max Weinreich, *Hitler's Professors: The Part of Scholarship in Germany's Crimes against the Jewish People* (New York: Yiddish Scientific Institute—YIVO, 1946), p. 40.
31. *Ibid.*, pp. 30–31, original italics.
32. *Ibid.*, p. 33.
33. *Ibid.*, p. 34.
34. Amitai Etzioni, *Genetic Fix, The Next Technological Revolution* (New York: Harper Colophon Books, 1975), p. 111.
35. "Biomedical Ethics and the Shadow of Nazism," Special Supplement in *The Hastings Center Report,* vol. 6, no. 4 (August 1976), pp. 1–18.
36. Piel, p. 44.

3

The Roles of Scientists in the Historic Cases

Seventeenth Century Italy

The behavior of Galileo's scientific peers stands as a contradiction of the assumption that a scientist's principal motivation is the pursuit and protection of truth. From the time that several scientists refused to look through Galileo's telescope in 1610 through his abjuration twenty-three years later, Galileo's cause was allowed to languish by most of his scientific peers. Even before the church announced disapproval of Copernican teaching in 1616, when support could have been given with relative impunity from church harassment,[1] Monsignor Dini had written to a friend that "many Jesuits are secretly of the same opinion, though they remain silent."[2]

Others cite the "incredible passiveness" of the mathematicians and scientists of the Jesuits' Collegio Romano. "Obedience and preoccupation with 'scandal' had been bred into the bones of the [Jesuit] Order, from Bellarmine down, to such an extent that the intellectual reflexes were dead."[3] The Jesuits, whose members included most of the mathematicians, physicists, and astronomers of the time, cared more about maintaining church orthodoxy and political advantage than speaking the truth as they knew it.[4]

Karl von Gebler attributed the Jesuits' hostility to insecurity about retaining their monopoly of science. Already shaken by the Reformation, the Jesuits were now confronted by Galileo the layman teaching "the astonished world truths before which the whole edifice of scholastic sophistry must fall to the ground." In consequence "the reformers of science appeared just as dangerous to them as those of religion."[5] In the end, writes Gebler, the Jesuits were responsible for inspiring the pope to believe that Galileo's writings were "eminently dangerous to the Church, more dangerous and abhorrent even than the writings of Luther and Calvin."[6]

The Galilean affair thus sets into relief two features that recur in our examination of the other historic cases: the embrace of politically imposed truth by several eminent scientists and the silent acquiescence of most others. While the pope and his inquisitorial court were politically obtuse and devoid of human charity, they were scientifically illiterate. Their actions may have been born more of naiveté than malice. But even this thread of mitigation must be denied the Jesuit mathematicians and astronomers. They knew better. If they had publicly linked their views to Galileo's, they might not only have saved Galileo from misfortune, but the church from an historical absurdity.

Brodrick excuses the Jesuits who "in their secret hearts . . . may very well have inclined to Galileo's conclusion." He offers a familiar justification. It is one of the sad links that connects the silent scientists in the Galilean, Lysenko, and Nazi episodes. The Jesuits were, after all, "under authority and could not teach anything they liked."[7]

The Soviet Union

Most Soviet scientists said nothing about Lysenko during the 1930s and 1940s. Their silence carried the weight of acquiescence, whatever their private thoughts. When criticism was officially acceptable, few scientists rose to oppose a pseudoscientific demagogue. Finally, in 1948, it was too late. Even scientists who had defended their disciplines against Lysenkoism had to recant. The forum in which these few stalwarts were heard, their initial protests, their professed loyalty to the political system and its ideology, their ultimate abjuration—all were reminiscent of Galileo before the church three hundred years earlier.

The session of the Lenin Academy of Agricultural Sciences that was held in August 1948 was no less extraordinary than the inquisitorial court that met in 1633. In 1948 the "true" nature of biology was established by political fiat. No questions, criticisms, or discussions were henceforth permissible if they challenged the principles announced at the session. The final authority was to be Lysenko. Lysenko did not announce his endorsement by the party's Central Committee until the end of the session, so that during the first few days, the handful of scientists willing to speak for classical genetics did not flinch.

Their speeches were respectful of the opposition, couched in praise of the Soviet system and dialectical materialism, yet unmistakably critical of Lysenko. Their comments were made prior to Lysenko's shattering announcement that the party approved only his approach. Perhaps the most poignant, painful episode in the thirty-five years of the Lysenko controversy occurred shortly after Lysenko's announcement near the end of the session. Several who earlier spoke for classical genetics recanted and declared support for the party decision. Excerpts from the earlier comments by S. I. Alikhanian and P. M. Zhukovsky show their commitment to classical genetics.

> *Alikhanian:* What would be the correct methodological approach to the problem of the gene from the standpoint of experimental genetics? The gene is an objective reality, a material unit in the living cell. Our task therefore is to estimate correctly the part played by the gene in the life of the cell and to give a proper materialistic explanation for the data that our science has accumulated. We cannot reject the sound and beneficial basis of genetics and disregard the facts obtained by science because this or that scientist has made reactionary statements. . . .
>
> The attacks on the view that genes exist reminds me of the earlier denials of the existence of the atom. Although the atom itself has never been seen, no one doubts its existence today. So it is with the chromosome. There were scientists who vigorously denied the reality of chromosomes. We claim that all plants and animals have specific chromosome complements which vary in number from two or three chromosome pairs to several hundred. . . .
>
> I therefore once again address myself to Trofim Denisovich, what is idealistic about that? If you consider the existence of the chromosomes and their affinity to the characters of the organism to be a fact, then why not study the structure of these chromosomes? A study of the structure of the chromosome (in your Institute of Genetics chromosomes are studied, aren't they?) discloses its het-

> erogeneity and linear dissimilarity. This has been definitely proved by experiment. I myself have confirmed it experimentally and checked my data cytologically.[8]

Zhukovsky outlined the scientific process of heredity, and he appeared firm in his commitment to genetics. When interrupted by Lysenko, he maintained his position uncompromisingly. His response was cautious, though tinged with condescension, as the following excerpts reveal.

> *Zhukovsky:* Our disagreements centre mainly around two questions: firstly, the chromosome theory of heredity, and secondly, the influence of external conditions. Trofim Denisovich Lysenko insists on a direct answer to these questions.
>
> As regards the chromosome theory of heredity. It would be deplorable if the entire group of geneticists that has been labelled Mendelist-Morganists were, from this rostrum, to renounce the chromosome theory of heredity. I do not intend to do that. . . .
>
> *Lysenko:* When I reply to the debate I will show you scores of plants, the parents of which were vegetative hybrids. Let any man who understands anything at all about hybridization say that they are not hybrids.
>
> *Zhukovsky:* Still, Trofim Denisovich, I beg of you to listen. I believe you about those plants, but I will call them mutations. What mutations are, I will tell you in a minute. Again I say that it is based on the chromosome theory. Mutation means a change in the chromosome that produces a genetic effect.
>
> Trofim Denisovich, you never use the term "mutation," you refuse to recognize it. But we do recognize it. And nature supplies the organic world with mutations almost without limit. What causes mutations? On this point, I am entirely with you, Academician Lysenko: environment, external conditions, cause mutations.
>
> You call it training. But that is beside the point. You refuse to admit that these mutations are caused by changes in the chromosome. That is where we disagree. It has gone so far that the very mention of the word "mutation" or "chromosome" frightens many people. . . .
>
> Our opponents never mention such terms as vitamins, hormones or viruses. I would advise, not you, Trofim Denisovich, your authority is sufficiently high, but your followers to study, for knowledge is light, while ignorance is darkness.[9]

Zhukovsky and the other anti-Lysenkoites demonstrate that, at least up to that moment, biologists could speak frankly in support of their discipline. Despite Lysenko's influence he never had the power to suppress his opponents. Had the opposition been more vocal during the thirties and forties, perhaps he would not have gained the position he was about to claim. Lysenko's statement toward the end of the session began with a terse announcement that decimated his opponents. His triumph was now absolute. From the transcript:

> Comrades, before I pass to my concluding remarks I consider it my duty to make the following statement.
>
> The question is asked in one of the notes handed to me, What is the attitude of the Central Committee of the Party to my report? I answer: The Central Committee of the Party examined my report and approved it.
>
> (*Stormy applause. Ovation. All rise.*)[10]

The following day a few scientists, including Alikhanian and Zhukovsky, asked to address the Academy. Their renunciations of earlier defenses of genetics were terrible echoes of Galileo's recantation. Once again, political authority had mandated scientific truth; opposition was no longer permissible.

> *Alikhanian:* When I leave this session, the first thing I must do is to review not only my attitude toward the new Michurinian science, but my entire earlier activity. I call upon my comrades to do likewise. . . .
>
> I shall strive to have my comrades not waste their experience and knowledge in vain, not leave them in their laboratories, but apply them broadly to the needs of the national economy. This is not hard to do, provided we rid ourselves of the deadweight of useless, metaphysical concepts, and honestly go forward to the end in close fellowship with all the scientists of our country.
>
> And only in our country, the country with the most advanced and progressive world outlook, can the seedlings of the new scientific trend develop. And our place is with this new, progressive trend. I, for my part, categorically declare to my comrades that I shall henceforth contend with those whose views I used to share and who do not understand this and do not take the way of the Michurinian trend.[11]

> *Zhukovsky:* There are moments in a man's life, especially in our historic days, which are to him of profound and crucial moral and political significance. This is what I experienced yesterday and today. The speech I made the day before yesterday was an unhappy one; it was the last of my speeches against Michurin, as it is said here to have been, although I personally have never

> before spoken in opposition to Michurin's teachings. At the same time, it was my last speech from an incorrect biological and ideological standpoint.
>
> The speech I made the day before yesterday, at the time when the Central Committee of the Party had drawn a dividing line between the two trends in biological science, was unworthy of a member of the Communist Party and of a Soviet scientist.
>
> I admit that the position I held was wrong. . . .
>
> The exceptional unity displayed by the members and guests at this session, the demonstration of the power of this unity and the bonds with the people, and, on the contrary, the demonstration of the weakness of the opponents are to me so obvious, that I declare that I shall fight—and there are times when I can fight—for the Michurinian biological science.[12]

The silence that had long characterized the response of most scientists to Lysenko's travesties was now officially mandated. From this time until after Stalin's death in 1953, biological science was whatever Lysenko said it was.

The scientist, by his inclination and training, is commonly assumed to seek truth based on evidence. As well as anyone, he should be able to separate myth from fact, intuition from evidence, chicanery from science. Moreover, scientists are professional experts who know more about their disciplines than nonscientists. But on the evidence in the Lysenko affair, most biological scientists did not protest assaults on truth as they surely understood it. A few opposed Lysenko forthrightly, while others joined his cause as he seemed to be gaining favor with the state.

If one expects that people who have bright minds, specialty training, and international eminence will, on the strength of these characteristics, reject politically condoned quackery, the Lysenko affair stands in contradiction. When encouraged by the political system, quackery prevailed and "good" scientists deferred to politically imposed scientific truth. Nowhere was this phenomenon more evident than among the scientists who served under the Third Reich.

Nazi Germany—The Physicists

Nazi doctrine about race affected every scientific discipline. Its application to physics and physicists was typical.

An odd assortment of adjectives began to appear before the word "physics" in the literature of the time. The descriptions fell into two categories: good, like "Nordic" physics, "Aryan" physics, "German" physics; and bad, like "modern" physics, "international" physics, "Jewish" physics. Proponents of these categories were not merely propagandists who lacked understanding of science and its methods, nor were they all second-rate scientists suffering from professional frustrations. They came from among the best of Germany's brilliant assemblage of physicists, Philipp Lenard, who had won the Nobel Prize before World War I, issued a statement in 1924 that enthusiastically supported Hitler and the Nazi party. Co-signed by Johannes Stark, who had also won the Nobel Prize in physics in 1919, the statement said:

> As recognized natural scientists we should like herewith to announce in conformity with our innermost feeling that in Hitler and his comrades we discern the same spirit which we always looked for, strove upward, developed out of ourselves in our work that it might be deep going and successful; the spirit of clarity without residue, of honesty toward the outer world, simultaneously of inner unity, the spirit which hates any compromise work because of its insincerity. This, however, is exactly the spirit which we earlier recognized and advanced in the great scholars of the past, in Galileo, Kepler, Newton, Faraday. We admire and adore it likewise in Hitler, Ludendorf, Pohner and their comrades; we recognize in them our nearest relatives in spirit. . . . [Hitler] and his comrades in struggle appear to us like God's gifts out of a time that has long past, in which races still were purer, men were still greater, minds less deceived.[13]

Lenard believed further that "the Jew conspicuously lacks understanding for the truth . . . being in this respect in contrast to the Aryan research scientist with his careful and serious will to truth. . . . Jewish physics is thus a phantom and a phenomenon of degeneration of fundamental German physics."[14]

In 1935 Stark elaborated on the subject:

> Jewish physics which . . . came into being during the last three decades and has been made and publicized both by Jews and by their non-Jewish pupils and imitators appropriately found its high priest in a Jew, in Einstein. Jewish advertising wanted to make

> him the greatest natural scientist of all times. Einstein's theories of relativity, however, in essence were nothing but a heaping of artificial formulae on the basis of arbitrary definitions and transformations of the space and time coordinates. . . . Jewish formalism in natural science is to be rejected by all means.[15]

Following suit were Professor Ludwig Bieberback of the University of Berlin, who called Einstein "an alien mountebank,"[16] and Professor Rudolphe Tomaschek, director of the Institute of Physics at Dresden, who wrote: "Modern physics is an instrument of [world] Jewry for the destruction of Nordic science. . . . True physics is the creation of the German spirit."[17] Professor Wilhelm Mueller, of the Technical College of Aachen, believed there was a worldwide Jewish plot to pollute science and destroy civilization. Einstein's theory, according to Mueller, was "directed from beginning to end toward the goal of transforming the living—that is, the non-Jewish—world of living essence, born from a mother earth and bound up with blood, and bewitching it into spectral abstraction in which all individual differences of peoples and nations, and all inner limits of the races, are lost in unreality."[18]

The willingness of German physicists to embrace Nazi gibberish is tragically demonstrated by one of the great scientists of the twentieth century, Max Planck. Although his sympathy for Nazism later waned, at first he appeared pleased to have Hitler speak for German nationalism. In May 1933, Planck presided over the twenty-second annual meeting of the Kaiser Wilhelm Society for the Advancement of the Sciences. Einstein had just resigned from the society under Nazi pressure, and Planck, formerly a friend as well as professional colleague, did not come to his support. Rather, Planck endorsed and read to the gathering the following message to Chancellor Hitler:

> The Kaiser Wilhelm Society for the Advancement of the Sciences begs leave to tender reverential greetings to the Chancellor, and its solemn pledge that German science is also ready to cooperate joyously in the reconstruction of the new national state.[19]

Most German physicists refrained from outrageous public declamations about German physics or Jewish physics. A few resigned from their positions in protest, even though not directly pressured by the Nazis for racial reasons. But the overwhelming majority acquiesced without objection.

Alan Beyerchen recounts that the German physicists were divided largely into two camps. One, led by Lenard and Stark, embraced an "Aryan" physics that denigrated concepts like relativity as false contrivances of the "Jewish spirit."[20] The other group tried to pursue their field apart from ideology. Thus Planck, Heisenberg, Finkelnburg, Ramsauer, and others who supported modern concepts argued that *theory* was not a Jewish but a German characteristic.[21] It was theory and its utility that they were concerned about, not the demise of the theory's creator if he happened to be Jewish. Only an exceptional few, like Otto Hahn and Max von Laue, refused to compromise and tried to help fellow physicists who were targets of the regime.[22]

The traditional physicists capitalized on intraparty factionalism and on the eventual need for their services to the war effort. They gained primacy over the "Aryan" physicists in the early forties.[23] Ironically, by employing their more rational physics, they were able to contribute to the Nazi military effort more effectively than the "Aryan" physicists ever could.

The Medical Experiments

Among the most grotesque manifestations of Nazi racial science were the infamous medical experiments performed by German doctors. Although the experiments were not themselves expressions of scientific truth, the use of involuntary "inferior" humans as subjects was a consequence of a perverted science. In acceding to the regime's dogma, members of the medical profession abandoned their pledge to relieve pain and suffering. They became agents of agony and death.

Experimentation on involuntary subjects began in 1939. While conducted in secret, knowledge about the experiments seeped through much of the German medical establishment, and the results were published in journals and reported at conferences. The experiments covered a variety of areas. Those elaborated on at the Nuremberg trials after World War II included high-altitude experiments, freezing experiments, malaria experiments, mustard gas experiments, sulfanilamide experiments, experiments on bone, muscle, and nerve regeneration, bone transplantations, sea water experiments, epidemic jaundice experiments, typhus and other vaccine experiments, experiments with poison, incendiary bomb experiments, phlegmon experiments, polygal experiments, gas edema (phenol) experiments, and experiments for mass sterilization. Other enterprises in the name of science included the collection of Jewish skeletons (for which people were carefully murdered in order to protect their skeletal structure), a project to kill tubercular Polish nationals, and euthanasia.[24]

An extract from the Nuremberg trial testimony of Walter Neff, an inmate assistant to Dr. Sigmund Rascher who conducted high-altitude experiments at the Dachau concentration camp, reveals the manner of selecting experimental subjects. The prosecutor was James M. McHaney.

Q. And you state that between 22 February 1942 and the end of June, or the beginning of July 1942, approximately 180 to 200 concentration camp inmates were experimented on?

A. Yes.

Q. What nationalities were the experimental subjects?

A. I cannot say that with certainty but I think that approximately all nations were represented there; that is, all nations that were in the camp, mostly Russians, Poles, Germans, and Jews belonging to any nation. I do not remember any other nationalities being represented there.

Q. Were any of these experimental subjects prisoners of war?

A. Yes.

Q. What nationalities were they? Do you recall?

A. They were Russians.

Q. Now, will you tell the Tribunal how these experimental subjects were selected?

A. The experimental subjects who had to be subjected to severe experiments, experiments that would end in death, were requested by Rascher from the camp administration and then furnished by the SS; however, this procedure differed with the so-called series of experiments and a number of other experiments. For those experiments, the people were brought into the experimental station straight from the camp, that is, from the blocks.

Q. Now, did they, to your knowledge, make any effort in the camp to secure volunteers for these experiments?

A. There were certain volunteers for these experiments. That was because Rascher promised certain persons that they would be released from the camp if they underwent these experiments. He sometimes promised them that they would be detailed to more favorable work.

Q. Now, about how many of such volunteers would you say there were for the high-altitude experiments?

A. I do not know the exact number. It was not very high; approximately 10 inmates volunteered for that purpose.

Q. Did these volunteers come one at a time, or did they come in a body, or just how did they present themselves to the experimental stations?

A. Rascher moved around the camp quite a lot and on that occasion the inmates spoke to him.

Q. In other words, the camp officials and Rascher and Romberg made no effort to find volunteers, did they?

A. I don't know, but I should not think so. I should not think that they made great efforts to get volunteers.

Q. Now, other than these approximately 10 persons who you state presented themselves as volunteers, were all the rest of the experimental subjects simply picked out and brought in and experimented on?

A. Yes.

. . . .

Q. Witness, were any Jews experimented on in these high-altitude experiments?

A. Yes.

Q. Now, tell the Tribunal approximately how many prisoners were killed during the course of the high-altitude experiments?

A. During the high-altitude experiments 70 to 80 persons were killed.

Q. Did they experiment on prisoners other than those condemned to death?

A. Yes.

Q. Were any of those prisoners who had not been condemned to death killed during the course of the high-altitude experiments?

A. Yes.

Q. Do you have any idea how many may have been killed?

A. There could have been approximately 40 persons.

Q. That is, 40 persons were killed, who had not been condemned to death, out of a total of 70, did you say?

A. Yes.[25]

Of the more than 3,200 human subjects used in the various experiments cited at the Nuremberg trials, about 1,100 died as a result, while survivors commonly suffered permanent physical and mental injury.[26] "In every one of the experiments the subjects experienced extreme pain or torture, and in most of them they suffered permanent injury, mutilation, or death, either as a direct result of the experiments or because of lack of adequate follow-up care," said the Nuremberg chief prosecutor.[27]

Not included among the victims were 10,000 tubercular Polish nationals killed by the Nazis, hundreds of thousands murdered in the name of euthanasia, and millions exterminated for racial reasons based on Nazi scientific truth. Nor can the number of doctors who participated in these activities be determined with precision. It can be said with certainty that many physicians were actively involved and that many more knew and did not object. Dr. Andrew C. Ivy, vice president of the University of Illinois, served as consultant at the Nuremberg trials at the request of the United States government upon recommendation of the American Medical Association. He discovered that several hundred physicians participated or had firsthand knowledge of the experiments. General awareness went beyond the medical community, but ultimate responsibility, according to Ivy, had to rest upon the "bulk of the German medical profession" because it did not protest.[28]

Despite professed intentions to keep the experiments secret, too many people knew about them to imagine that word did not spread. Leaders of the German medical profession—in government, in the military, in medical schools, in research institutions—were aware; some were performing them. There are inferences in the Nuremberg documents that an occasional doctor demurred from participation. But in the final, sad analysis, among the many hundreds of medical scientists who knew firsthand about the experiments, and the countless others who knew inferentially, virtually no one raised any objection.

WERNER FORSSMANN—A SKETCH

The reaction of Werner Forssmann, a distinguished medical scientist, could have been that of Galileo's or Lysenko's scientific peers. His memoirs are replete with self-justification and excuses for going along with the state and its version of scientific truth.

As a young doctor in 1929, Forssmann conducted pioneering work on heart catheterization. Soon after, swept by Hitler's promises of national stability, he joined the Nazi party. He remained a member while working as a medical administrator and surgeon throughout the Hitler years. His party activities did not preclude his receiving honors after the war, including the Nobel prize in 1956 for his earlier studies of the heart.

Forssmann apparently did not participate in any of the infamous medical experiments. Yet, even in his self-serving autobiography, Forssmann reveals his knowledge of "eugenic sterilizations" performed by many of his associates, including distinguished surgeons like Sauerbruch, Fromme, and Krohn. He denies having personally performed any, though he jokingly writes, "I must have cut more spermatic cords in preparing for prostate gland operations than many a surgeon."[29] Forssmann's ambivalence toward the issue is reflected in his calling the sterilization surgery "soul-destroying work," yet defending his associates' activities on three counts. First, sterilization was done on "mental patients." Second, the patients were "thoroughly examined by psychiatrists, and the indications clearly established." Third, it was a "duty imposed . . . by the state."[30] Not only did Forssmann refrain from suggesting to his associates that they refuse to perform "eugenic sterilizations" (nor does he italicize "eugenic"), but he completes his rationalization on the basis of duty to the state, as if it superseded all other obligations.

Forssmann later tells of his refusal of an offer, "which was in itself tempting, but circumstances made it totally unacceptable," to perform cardiological experiments at a sanatorium using "defenseless patients as guinea pigs."[31] While denying himself the indulgence of this "tempting" opportunity, Forssmann continued his work under the Third Reich outwardly oblivious to the medical horrors he knew were going on.

Forssmann was not unique. Countless doctors, nurses, orderlies, and other personnel had knowledge of the medical experiments. If most did not participate directly, did those who remained silent share responsibility? Should Forssmann at least have urged his good friend Fromme to refuse to perform "eugenic sterilizations"? Forssmann claimed that he was disenchanted with Hitler by 1934 and that he went into "internal emigration."[32] Nevertheless, he personally fared well throughout the Nazi years because he cooperated with the regime. As with so many other Germans who claimed "internal emigration" as their way of dealing with Hitler, his outward cooperation, in effect, meant support.

Forssmann's behavior was typical of the German doctors who knew better. He could have been one of the acquiescent Jesuit scientists in the 1600s, or a silent biologist in the Soviet Union during the 1930s and 40s. As will be suggested in the next chapter, he was what most of us are. Despite affronts to our sense of morality or truth, when speaking out might cause us loss of position or inconvenience, most of us remain silent. In the historic cases involving Galileo, Lysenko, and the Nazis, the silence of the scientists contributed to tragic consequences. Whether their responses were labeled "internal emigration" or "prudential acquiescence,"[33] their effect was to support politically imposed truths and perverse activity in the name of science.

Notes

1. Until 1616, as long as the Copernican system was "confined to the language of science" and was not supported as a contradiction of scripture, its discussion was "not only permitted, but encouraged." See Arthur Koestler, *The Sleepwalkers* (New York: Macmillan Co., 1959), p. 357.
2. Quoted in Ludovico Geymonat, *Galileo Galilei: A Biography and Inquiry into His Philosophy of Science* (New York: McGraw-Hill Book Co., 1965), p. 85.
3. Giorgio de Santillana, *The Crime of Galileo* (Chicago: University of Chicago Press, 1955), p. 143.
4. *Ibid.*, pp. 194, 204–5; Geymonat, pp. 75, 85.
5. Karl Von Gebler, *Galileo Galilei and the Roman Curia,* translated by Mrs. George Sturge (Merrick, N.Y.: Richwood Publishing Co., 1977), p. 154.
6. *Ibid.*, p. 162.
7. James Brodrick, *Galileo: The Man, His Work, His Misfortunes* (London: Geoffrey Chapman, 1964), p. 51.
8. *Proceedings of the Lenin Academy of Agricultural Sciences of the U.S.S.R., The Situation in Biological Science,* Session July 31–August 1, 1948, Verbatim Report (Moscow: Foreign Languages Publishing House, 1949), pp. 426–32.
9. *Ibid.*, pp. 456–64.
10. *Ibid.*, p. 605, original italics.
11. *Ibid.*, pp. 621–22.
12. *Ibid.*, pp. 618–19.
13. Max Weinreich, *Hitler's Professors: The Part of Scholarship in Germany's Crimes against the Jewish People* (New York: Yiddish Scientific Institute—YIVO, 1946), p. 12.

14. William L. Shirer, *The Rise and Fall of the Third Reich* (New York: Simon and Schuster, 1960), p. 251.
15. Weinreich, p. 12.
16. Shirer, p. 251.
17. *Ibid.*, p. 250.
18. *Ibid.*
19. Ronald W. Clark, *Einstein, The Life and Times* (New York: World Publishing Co., 1971), p. 470.
20. Alan D. Beyerchen, *Scientists under Hitler: Politics and the Physics Community in the Third Reich* (New Haven, Conn.: Yale University Press, 1977), p. 132.
21. *Ibid.*, pp. 184–85.
22. *Ibid.*, p. 208.
23. *Ibid.*, pp. 168–98.
24. *Trials of War Criminals before the Nuernberg Military Tribunals under Control Council Law No. 10*, "The Medical Case," Vols. 1 and 2, October 1946-April 1949 (Washington, D.C.: U.S. Government Printing Office), p. 56.
25. *Ibid.*, Vol. 1, pp. 68–69.
26. I have compiled these figures from the documentary evidence presented by the prosecution throughout the trial.
27. Telford Taylor in Alexander Mitserlich and Fred Mielke, *Doctors of Infamy, The Story of the Nazi Medical Crimes* (New York: Henry Schuman, 1949), p. xxv.
28. Andrew C. Ivy in *ibid.*, pp. x–xi.
29. Werner Forssmann, *Experiments on Myself, Memoirs of a Surgeon in Germany* (New York: St. Martin's Press, 1974), p. 179.
30. *Ibid.*
31. *Ibid.*, p. 241.
32. *Ibid.*, p. 161.
33. The roots of prudential acquiescence among scientists are traced to Descartes and Bacon in Joseph Haberer, *Politics and the Community of Science* (New York: Van Nostrand Reinhold Co., 1969), chap. 5.

4

Why Scientists Acquiesced to Politically Imposed Truth

This chapter inquires into the motivation behind scientific activity and how it might relate to a scientist's acceding to externally imposed scientific truth. If, as some claim, a scientist's fundamental motivation is to discover truth, then scientists' acquiescence to falsehoods mandated by the state seems a curious contradiction. Yet, as will be suggested in this chapter, scientists in the historic cases behaved as scientists today probably would under similar circumstances.

Five propositions are examined to explain the acquiescence by scientists in the historic cases: personality traits of individuals, fear of the consequences of dissent, desire to protect their disciplines, competition among scientists, and the human propensity to obey authority. Each may have a measure of validity, though when dealing with a scientist's response to his political system, the last assumes overriding importance. Consequently, the nature of a political system appears preeminent as a determinant of whether a state will impose truth and whether scientists in that society will acquiesce.

Personality

Several scholars have sought to explain a scientist's behavior purely in terms of his personality and his psychological orientations. In the context of this study, such explanations are inadequate, for they fail to account for the influence of the political environment.

Ian Mitroff, for example, categorizes scientists according to one of four modes, depending on their perception and evaluation of phenomena: sensation-thinking, sensation-feeling, intuition-thinking, and intuition-feeling.[1] He applies his typology to scientists that he interviewed, but he suggests that the scheme is universally applicable. Thus the perception-evaluation scale should be applicable to scientists in Galileo's Italy or in twentieth century Germany and Russia. Even if we were capable of placing those scientists into Mitroff's typology, the information is irrelevant to externally imposed scientific truths in their societies. It tells nothing about which scientists would likely embrace or reject perverted truths about their disciplines.

Drawing on Mitroff's study, George F. Kneller discusses three personality types among scientists: theorizers, empiricists, and intermediates.[2] His descriptions are no more informative than Mitroff's about the likelihood of responses to imposed truth. Similarly, Ann Roe's analysis of the results of Rorschach tests administered to scientists reveals nothing about how they might react to politically mandated truths.[3] Roger G. Krohn's inquiry among scientists, at first, appears more relevant when he seeks "the degree to which freedom for the individual or efficiency for the organization is considered more important in the administration of research."[4] Yet, even if the few scientists who favored organizational efficiency are receptive to involvement by the state, this would not necessarily mean they would acquiesce to a political imposition of scientific truth.

The same uncertainty arises from responses to a questionnaire by Marlan Blissett. Scientists overwhelmingly affirmed their "commitment to resist regulation or control of scientific research." Seventy-seven percent agreed that "the pursuit of science is best organized when as much freedom as possible is granted to all scientists." Only 15 percent disagreed.[5] But how the scientists would behave when faced with a mandate from outside is uncertain. One can hardly be optimistic, based on the behavior of scientists in the historic cases, and psychological and personality assessments alone are inadequate explanatory instruments.

Misperceptions about the Consequences of Dissent

A common theme among the three historic cases was the overestimation by scientists of the dangers that would attend their resistance. Since most scientists complied, questions about resistance must remain speculative. Nevertheless, opposition by others to political policies was, at times, successful. In Germany, for example, after December 1939 mentally deficient persons were ordered exterminated by Hitler. As knowledge of the euthanasia program became widespread, popular disapproval mounted. Efforts to stop the program were led not by scientists or doctors, whose professional colleagues were involved in the selection or extermination of victims, but by a few Protestant and Catholic clergy. Their disapproval was climaxed in August 1941, when the Bishop of Munster publicly denounced the murder of innocent sick people. Immediately afterward, fearing public unrest, the government halted the program and the leading opponents of the euthanasia program went unmolested.[6] While scientists remained silent about the regime's corruption of their disciplines, others demonstrated that opposition to Nazi policies could be effective.

The situation prompted Peter Hoffmann to single out scientists and other intellectuals during the Hitler years as guilty of "acquiescence, weakness, opportunism, delusion and error."[7] Joseph Haberer wrote that "more than any [other] institutional or professional community in Germany, the scientists disengaged themselves from the problem of their responsibility in the crisis which involved them."[8] He added:

> It is clear that German scientists at the time [that Hitler gained power], and afterward, overestimated the dangers involved, and tended to assume that one risked one's life by not complying with the explicit or implied wishes of the regime in minor matters. But men who took much stronger positions survived.[9]

In the Soviet Union during the 1930s and 40s, no segment of the population escaped Stalin's whim. Scientists, like others, were subjected to purges and terror, and the reluctance of Soviet scientists to criticize Lysenko was sometimes attributed to their fear of antagonizing a political leadership that ostensibly supported Lysenko. But based on a careful analysis of scientists who were harassed and others who were not, David Joravsky determined that the fear was unwarranted. Many Lysenkoites suffered arrest, and "most of the most vigorous opponents of Lysenkoism were not arrested," concluded Joravsky. "Any way one searches it, the public record simply will not support the common belief that the apparatus of terror consciously and consistently worked with the Lysenkoites to promote their cause."[10]

Even after Stalin's death, when criticism of Lysenkoism did not endanger their professional positions, relatively few scientists spoke out in opposition. Those who did included not only biologists, but physicists, chemists, and others. Thus, while scientists in many disciplines were obviously aware of Lysenko's fraudulent principles, the vast majority remained silent. They might have feared that Khrushchev's warmth toward Lysenko would jeopardize their positions, though such a judgment proved baseless.[11]

A misapprehension of the penalties attendant to resisting external authority also entrapped scientists during the Galilean incident. Some opposed Galileo merely to curry favor with the authorities, others because of personal dislike for Galileo.[12] Most remained silent out of habituated obedience to church authority. The Jesuit astronomers bore the greatest responsibility for Galileo's demise for, unlike others, they understood his work. Their commitment to church authority and fear of challenging it superseded their commitment to scientific truth.

> They were more than half-convinced that Galileo was right. . . . Galileo had plenty of information about their thought, and he kept expecting that, whatever their personal feelings, their duty to the Faith would cause them to interpose a word of advice. It was they, and no one else, whose obligation it was to prevent the Pope from making a fool of himself. But the vast apparatus of indoctrination and constriction that their Order had devised was now working to its own undoing. Following "like unto corpses" the corporate political will of their Society, they shut their eyes, their ears, and their minds. The power of discipline fed back into the complex steering machinery in a circuit of self-destruction.[13]

Accounts of all three cases indicate that the large majority of scientists overestimated the need to cooperate with their regime's perversion of science. Because of fear or habituation, they allowed scientific truth to be corrupted. They seemed little concerned that their acquiescence would confer a legitimacy on the imposed truth that the political authorities alone were not capable of doing. When the scientific experts went along with the regime's definition of scientific truth, no one was left to challenge its validity.

Salvaging Their Disciplines

Another proposed explanation to account for acquiescence among scientists is that by cooperating with the regime, other work could continue unhindered. Werner Forssmann speaks of how he and many of his associates engaged in medical practice and research under Hitler even though they privately were opposed to his policies.[14] By ignoring the regime's transgressions, they supposed, they could continue their "good" work.

The physicists Max Planck and Werner Heisenberg offered no resistance to Hitler ostensibly in order to save as much as they could of their disciplines from destruction by the regime.[15] "Loyalty to Germany," wrote Beyerchen about such scientists, "meant devotion to the traditions and institutions of German science and efforts to preserve them, even if these efforts entailed moral compromises through apparent—or real—cooperation with the Nazi state."[16]

The same rationalizations were used regarding the behavior of Soviet scientists during the rise of Lysenkoism. Zhores Medvedev, a former student of Peter M. Zhukovsky's, recalled that the famous botanist told him in 1948 that his announced support for Lysenko's biology was a temporary accommodation. He anticipated that Lysenkoism would not remain dominant for long. In the meantime, Zhukovsky wished to preserve as much traditional science as he could by retaining his academic position and supporting his students.[17]

Similarly, N. A. Maksimov, who had been an eminent plant physiologist, became an apologist for Lysenko in the 1940s. Some saw this as a rear-guard action in defense of traditional science. By inserting a few pages of Lysenkoism into his textbook, Maksimov was able to keep the remaining valuable portions on plant physiology in print. By retaining administrative power, he could protect scientists engaged in research on plant hormones and other areas attacked by Lysenko, at least until the party edict in 1948. Zhukovsky, Maksimov, and others "engaged in pliable defense of [their] science, letting the enemy in at points of greatest pressure to prevent the complete destruction of the whole enterprise."[18]

The desire to preserve scientific truth appeared to account for Galileo's behavior after he was first admonished by Cardinal Bellarmine. Following the Cardinal's warning in 1616 not to expound Copernican ideas, Galileo ceased to discuss the issue. Rather than risk condemnation, he became ostensibly chastened of Copernicanism, but his retreat was tactical. When he thought he would have the support of a sympathetic pope fifteen years later, he reasserted his conviction by writing the *Dialogue Concerning the Two Chief World Systems.* He had "emigrated internally" during those fifteen years to buy time for scientific truth—an act of preservation shared in spirit three hundred years later by Heisenberg and Maksimov.

Galileo's decision to reaffirm publicly his belief in Copernicanism by publishing the *Dialogue* in 1632 turned out to be a supreme misjudgment. He was stunned by the pope's opposition and by the lack of support from his peers. Nevertheless, his silence between 1616 and 1632 represented an attempt to preserve scientific truth until the time seemed propitious to promulgate it, a theme that threads through each of the historic cases. Scientists have cooperated with dictatorial regimes in the belief that this was the best way to preserve at least some of their science.

Competition

A fourth possible reason that scientists cooperated with the regimes in the three historic cases arises from a proposition by Robert K. Merton. He observes that many scientists share a single behavioral imperative—intense competition for discovery. They respond not only to an internal "insatiability of wants," but to peer group pressure to achieve more and more. "In this respect," writes Merton, "the behavior of scientists does not much vary, transcending differences of time and national culture."[19]

Warren O. Hagstrom studied the scientific community in the United States from the perspective suggested by Merton. He found that outright fraud was rare, yet "scientists may violate the norms of free communication in science in order to protect their competitive standing, and they may default on their obligation to present only thoroughly verified results in the interest of publishing quickly."[20] The intensity of competition is amusingly though graphically portrayed by James D. Watson about his and Francis Crick's race to discover the structure of DNA, the molecule of heredity. At times they were consumed by anxiety when they believed that Linus Pauling was nearer than they were to making the discovery. The success of others brought gloom, the failure of others joy and celebration.[21]

The intensity of competition among scientists should not be underestimated. Descriptions of research activity at times conjure images of a combat zone. When physicists at Stanford University learned in 1979 that a West German group had found evidence for the existence of the gluon before they did, they felt "humiliated," according to a report in *Science* magazine. "We could have been number one," said a Stanford physicist, who was quoted beneath the banner: "Physicists are bitter about defeat to West Germany."[22]

In a review of the current condition of medical research, a physician describes how "scientists *battle* to be the first to publish, in order to establish priority with a finding."[23] Elsewhere, a report characterizes rival groups engaged in recombinant DNA research as involved in an angry race, and the competition "has created such animosity between some researchers that they no longer speak to each other and has led to threats of patent-infringement lawsuits."[24]

David Joravsky concludes that almost all scientists are "single-minded devotees of winning." He suggests cynically that a scientist's social and moral worth depends on "his capacity to beat the competition, to win, whether fame for himself or wars for his country, or both together."[25]

In view of the intense drive for recognition among scientists, their acquiescence to political imposition becomes more understandable if, in their perception, acquiescence might enhance their status. This is not to say that in order to gain recognition most scientists would abandon all scruples and embrace whatever perversion a political regime proposes, but it does point to a characteristic of the field and of its participants that separates it from others. Science, as Merton says, creates an environment in which the quest for acclaim may become a "driving lust."[26]

Of course, if one actively supported a perverted science, one's glory would remain within the borders of the warped society that legitimized the perversion. Nevertheless, silence and acquiescence might seem the safest route to continuing one's work in areas not directly affected by the perversion. As scientists rationalized their cooperation in the name of preservation of their science, their motivation might have involved their drive for personal advancement. In this light, the ethos of science fosters values and attitudes that contradict commonly held assumptions.

Scientists are better educated than most others. Their training might be expected to help them make objective and rational judgments and to protect the truths of their disciplines. Scientists *should* be least likely to allow the state to pervert scientific truth. Yet when external political authority has mandated scientific dogma, most scientists have not been courageous defenders of truth. As demonstrated in the historic cases, contrary to Bronowski's characterization, scientists did succumb to "that grey appeal to expediency."

Politics, Authority, and Evil

In seeking full understanding of a scientist's response to imposed dogma, one must look beyond the scientists. If the nature of a political system is an essential determinant of whether scientific truth will be imposed from outside the science, the political leaders are integral to the question. Although the following commentary about political leaders is based on the Nazi experience, its lessons are applicable to the Galilean and Lysenko cases as well.

Hannah Arendt in her report on the trial of Adolf Eichmann raised a controversial question about the character of the Nazi leaders. It turned on whether the Nazis were invidious, rapacious individuals, or whether they were essentially normal, each performing his job under the legitimate authority of the state. Eichmann was tried in 1961 for committing crimes related to his activities during the Nazi period, including crimes against humanity and the Jewish people. He had been in charge of organizing the forced emigration of Jews from Germany in 1938, and later was responsible for their transportation to the extermination camps. Gideon Hausner, chief prosecutor at the trial in Jerusalem, characterized Eichmann as "a dangerous, perverted, sadistic personality."[27] Hausner's understanding of Eichmann's psyche implicitly extended to other Nazis, especially to participants in the commission of atrocities. Hausner's views were rejected by Hannah Arendt in her book, *Eichmann in Jerusalem,* significantly subtitled *A Report on the Banality of Evil.*[28]

Arendt held that most Nazi leaders were not psychopathic. "The trouble with Eichmann was precisely that so many were like him, and that the many were neither perverted nor sadistic, that they were, and still are, terribly and terrifyingly normal."[29] They were, as they claimed, merely performing the tasks designated appropriate and legitimate by their political system, according to Arendt. In consequence, their roles could have been filled by many who would have acted similarly.

Contradictory positions have subsequently been urged. In 1975 Florence Miale and Michael Selzer published their interpretations of Rorschach tests that had been taken by Nazi leaders during the Nuremberg trials. In conjunction with a review of the Nazis' personal histories, the authors concluded that virtually all of the Nuremberg defendants whose tests they examined were psychopathic.[30] They also reviewed the Rorschach record of Adolf Eichmann and dismissed Arendt's "claim regarding the normality of the Nazi personality as represented by Adolf Eichmann."[31] "The Nazis," concluded the authors, "were not psychologically normal or healthy individuals."[32]

Soon after the appearance of the Miale-Selzer book, another scholar determined from his review of the literature that the Nazis were neither sadists nor deranged. Askenasy concluded that "they were essentially normal, ordinary men." Even more chilling: "Those Nazis and these Americans—which is to say you and I—for all our superficial differences such as time and place, are psychologically speaking, interchangeable."[33]

Askenasy relies on the reports of psychiatrists and psychologists who examined Nazi leaders at Nuremberg. His proposition of interchangeability largely derives from experiments conducted by Stanley Milgram, which offer powerful evidence that most people acquiesce to the dictation of political authority. The experiments provide the most useful explanatory matrix toward understanding a politically induced perversion of science. They confirm that the central element is found in the nature of the authority—what the authority tells its constituents to do.

Transposed to a larger scale, the results of Milgram's work are a reaffirmation that the likelihood of the occurrence of imposed truth can best be gauged, not from the nature of science or scientists, but from the character of political systems.

MILGRAM'S EXPERIMENTS

Stanley Milgram's *Obedience to Authority* demonstrates the extent to which people are willing to inflict pain on others when told to do so by authority. It integrates, on a micro-level, the three features that have undergirded our discussion: imposed scientific truth (the inflicting of pain in Milgram's experiments), political authority (the experiment leader), and acquiescent scientists (subjects obedient to authority). Milgram described his experiments:

> Two people come to a psychology laboratory to take part in a study of memory and learning. One of them is designated as a "teacher" and the other a "learner." The experimenter explains that the study is concerned with the effects of punishment on learning. The learner is conducted into a room, seated in a chair, his arms strapped to prevent excessive movement, and an electrode attached to his wrist. He is told that he is to learn a list of word pairs; whenever he makes an error, he will receive electric shocks of increasing intensity.
>
> The real focus of the experiment is the teacher. After watching the learner being strapped into place, he is taken into the main experimental room and seated before an impressive shock generator. Its main feature is a horizontal line of thirty switches, ranging from 15 volts to 450 volts, in 15-volt increments. There are also

> verbal designations which range from SLIGHT SHOCK to DANGER—SEVERE SHOCK. The teacher is told that he is to administer the learning test to the man in the other room. When the learner responds correctly, the teacher moves on to the next item; when the other man gives an incorrect answer, the teacher is to give him an electric shock. He is to start at the lowest shock level (15 volts) and to increase the level each time the man makes an error, going through 30 volts, 45 volts, and so on.
>
> The "teacher" is a genuinely naive subject who has come to the laboratory to participate in an experiment. The learner, or victim, is an actor who actually receives no shock at all. The point of the experiment is to see how far a person will proceed in a concrete and measurable situation in which he is ordered to inflict increasing pain on a protesting victim. At what point will the subject refuse to obey the experimenter?
>
> Conflict arises when the man receiving the shock begins to indicate that he is experiencing discomfort. At 75 volts, the "learner" grunts. At 120 volts he complains verbally; at 150 he demands to be released from the experiment. His protests continue as the shocks escalate, growing increasingly vehement and emotional. At 285 volts his response can only be described as an agonized scream.[34]

Approximately two-thirds of the subjects continued to the last shock on the generator. In variations of the experiment, Milgram found that fewer subjects complied as their proximity to the "learner" was increased. But even when the "teacher" had to press the "learner's" hand onto a shock plate while listening to his agonized pleas to stop, 30 percent continued up the generator to the end.

Milgram comments:

> A reader's initial reaction to the experiment may be to wonder why anyone in his right mind would administer even the first shocks. Would he not simply refuse and walk out of the laboratory? But the fact is that no one ever does. Since the subject has come to the laboratory to aid the experimenter, he is quite willing to start off with the procedure. There is nothing very extraordinary in this, particularly since the person who is to receive the shocks seems initially cooperative, if somewhat apprehensive. What is surprising is how far ordinary individuals will go in complying with the experimenter's instructions. Indeed, the results of the experiment are both surprising and dismaying. Despite the fact

that many subjects experience stress, despite the fact that many protest to the experimenter, a substantial proportion continue to the last shock on the generator.

> Many subjects will obey the experimenter no matter how vehement the pleading of the person being shocked, no matter how painful the shock seems to be, and no matter how much the victim pleads to be let out. This was seen time and again in our studies and has been observed in several universities where the experiment was repeated. It is the extreme willingness of adults to go to almost any lengths on the command of an authority that constitutes the chief finding of the study and the fact most urgently demanding explanation.[35]

Milgram's work appears to support Hannah Arendt's thesis about the nature of evil. "The ordinary person," says Milgram, "who shocked the victim did so out of a sense of obligation—a conception of his duties as a subject—and not from any particularly aggressive tendencies." Milgram concluded that the most fundamental lesson of his study was that "ordinary people, simply doing their jobs, and without any particular hostility on their part, can become agents in a terrible destructive process."[36] Milgram despairs that the human species ultimately will not survive this "fatal flaw nature has designed into us."[37]

It would seem that few could find comfort in the results of the Milgram experiments. Curiously, Barrington Moore has reinterpreted Milgram's work in an optimistic light based on the diminished compliance of the subjects under certain conditions. He is encouraged by the fact that fewer subjects complied as they were placed nearer to the victim, when the authority figure (the experiment leader) left the room, or when authority was otherwise made less direct.[38]

Moore's efforts to soften Milgram's implications are strained. To derive solace from the conditions that Moore does is to ignore the greater inference. It is alarming that two-thirds of a population would obey an order to inflict severe pain on helpless victims when noncompliance would draw absolutely no penalties. It is hardly less alarming that one-third would physically force a victim's hand onto a metal plate in order to administer SEVERE and DANGEROUS shocks to an innocent person painfully pleading to be released.

As for the decrease in compliance when the authority is more ambiguous, this supports Arendt's and Milgram's argument. Most people are not sadists. Left to themselves, as Moore says, they might well be humane and rational. But under conditions when they are not left to themselves, when they are directed by authority, they are capable of perpetrating heinous acts.[39]

One may infer from this how the overwhelming majority of people becomes pliant under politically repressive regimes. Sanctions for defying a political regime are potentially enormous, while those for defying the experimenter in the Milgram study were nil. If most subjects obeyed Milgram's experimenter, their obedience to political authority should not be surprising. The failure of scientists to resist the imposition of scientific truth appears bland beside Eichmann's activity or even that of the subjects in the Milgram study. Scientists share with most people a disinclination to oppose political authority. The character of their field, insofar as it fosters competition and thirst for recognition, may enhance the likelihood of their acquiescence.

A political system is comprised of many elements—economic relationships, political and social values, hierarchical structure. It is an awesome blend of elements that largely shapes the norms of citizens' behavior. Some citizens might overtly denounce what appears to them to be a politically mandated falsehood, but when confronted by pressure from the state, few seem able to summon the will to stand in opposition. Most are unwilling to denounce shams that are enunciated by political authority in the guise of scientific truth. As exemplified by the historic cases, scientists played secondary roles in the imposition or prevention of perverted science. When political authorities imposed corrupt policies or mandated "truths," scientists offered as little resistance as other groups, sometimes less.

Notes

1. Ian I. Mitroff, *The Subjective Side of Science* (New York: Elsevier Publishing Co., 1974), pp. 167–69.
2. George F. Kneller, *Science as a Human Endeavor* (New York: Columbia University Press, 1978), pp. 161–63.

3. Anne Roe, *The Making of a Scientist* (New York: Dodd, Mead and Co., 1953), pp. 194–213.

4. Roger G. Krohn, *The Social Shaping of Science* (Westport, Conn.: Greenwood Publishing Corp., 1971), p. 118.

5. Marlan Blissett, *Politics in Science* (Boston: Little, Brown and Co., 1972), pp. 67, 73.

6. Gunter Lewy, *The Catholic Church and Nazi Germany* (New York: McGraw-Hill Book Co., 1964), pp. 264–66. As Lewy shows, however, opposition to the euthanasia program by church leaders was unique. They remained silent about other Nazi policies, including official racism and extermination of Jews.

7. Peter Hoffmann, *The History of the German Resistance 1933–1945* (Cambridge, Mass.: The M.I.T. Press, 1977), pp. 12–13.

8. Joseph Haberer, *Politics and the Community of Science* (New York: Van Nostrand Reinhold Co., 1969), p. 141.

9. *Ibid.*, p. 138. Also see Alan D. Beyerchen, *Scientists under Hitler* (New Haven, Conn.: Yale University Press, 1977), p. 208.

10. David Joravsky, *The Lysenko Affair* (Cambridge, Mass.: Harvard University Press, 1970), p. 122.

11. Zhores A. Medvedev, *Soviet Science* (New York: W. W. Norton and Co., 1978), p. 90. Lysenko's influence on biology, though attenuated, continued through 1964 when Khrushchev was dismissed from power. But during Khrushchev's tenure, traditional biology regained legitimacy and opposition to Lysenkoism was not considered heretical. In fact, Lysenko has never been officially discredited, and in the 1970s a few Soviet biologists were still espousing some of his theories. *Ibid.*, pp. 226–31.

12. Giorgio de Santillana, *The Crime of Galileo* (Chicago: University of Chicago Press, 1955), p. 290.

13. *Ibid.*, p. 204–5.

14. Werner Forssmann, *Experiments on Myself, Memoirs of a Surgeon in Germany* (New York: St. Martin's Press, 1974), p. 161.

15. Beyerchen, pp. 63–64.

16. *Ibid.*, p. 64.

17. Medvedev, pp. 144, ff.

18. Joravsky, p. 199.

19. Robert K. Merton, "Behavior Patterns of Scientists," *American Scientist*, vol. 57, no. 1 (1969), p. 7; and "The Matthew Effect in Science," *Science*, vol. 159, no. 3810 (January 5, 1968), p. 57.

20. Warren O. Hagstrom, *The Scientific Community* (New York: Basic Books, Inc., 1965), p. 100. Others suggest that fraud among scientists has been more common than is generally assumed. See William Broad and Nicholas Wade, *Betrayers of the Truth* (New York: Simon and Schuster, 1982).

21. James D. Watson, *The Double Helix* (New York: Mentor, New American Library, 1968), pp. 103–4.

22. William J. Broad, "Stanford Fails to Score in 'Gluon Race,' " *Science*, vol. 205, no. 4413 (September 29, 1979), p. 1361.

23. Lawrence K. Altman, "The Doctor's World: How Honest is Medical Research?" *The New York Times*, August 5, 1980, p. C-3 (my italics).

24. William Stockton, "On the Brink of Altering Life," *The New York Times Magazine,* February 17, 1980, p. 18.

25. David Joravsky, "The Scientist as Conformist," *The New York Review of Books,* October 12, 1978, p. 37.

26. Merton, "Behavior Patterns of Scientists," p. 19.

27. Gideon Hausner, "Eichmann and His Trial," *The Saturday Evening Post,* November 3, 1962, p. 20.

28. Hannah Arendt, *Eichmann in Jerusalem, A Report on the Banality of Evil* (New York: The Viking Press, 1970).

29. *Ibid.,* p. 276.

30. Florence R. Miale and Michael Selzer, *The Nuremberg Mind, The Psychology of the Nazi Leaders* (New York: Quadrangle/The New York Times Book Co., 1975), p. 280.

31. *Ibid.,* pp. 7, 289–92.

32. *Ibid.,* p. 287.

33. Hans Askenasy, *Are We All Nazis?* (Secaucus, N.J.: Lyle Stuart, Inc., 1978), p. 49. Similarly Henry V. Dicks, a psychiatrist who interviewed eight former SS men convicted of "brutal mass murder of defenceless persons," concluded that they were not " 'insane' or uncontrollable people, in any generally understood clinical sense." *Licensed Mass Murder: A Socio-psychological Study of Some SS Killers* (New York: Basic Books, 1972), p. 230.

34. Stanley Milgram, *Obedience to Authority* (New York: Harper Colophon Books, 1975), pp. 3–4.

35. *Ibid.,* pp. 4–5.

36. *Ibid.,* p. 6.

37. *Ibid.,* p. 188.

38. *Ibid.,* chapters 8 and 9; Barrington Moore, Jr., *Injustice, The Social Bases of Obedience and Revolt* (White Plains, N.Y.: M. E. Sharpe, 1978), pp. 98–101.

39. Milgram's experiments have been criticized for employing subjects without their informed consent. In response, Milgram recalls that most experiments in social psychology would be impossible to conduct without a degree of staging, and that unlike some biomedical research they present no risks to the subjects. Milgram holds that "it is extremely important to make a distinction between biomedical interventions and those that are of a purely psychological character, particularly the type of experiment I have been discussing. Intervention at the biological level *prima facie* places a subject 'at risk.' The ingestion of a minute dose of a chemical or the infliction of a tiny surgical incision has the potential to traumatize a subject. In contrast, in all of the social psychology experiments that have been carried out, there is no demonstrated case of resulting trauma. And there is no evidence whatsoever that when an individual makes a choice in a laboratory situation—even the difficult choices posed by the conformity or obedience experiments—any trauma, injury, or diminution of well-being results." Stanley Milgram, "Subject Reaction: The Neglected Factor in the Ethics of Experimentation," *The Hastings Center Report,* vol. 7, no. 5 (October 1977), p. 22. Whether or not one objects to staged experiments, the inferences in Milgram's work about response to authority remain unaffected.

5

Science and Politics: The Historic Cases and the United States

This chapter reviews the conditions of science and politics in the societies of the historic cases and compares them to those in the United States. If scientific inquiry has long been handicapped in a community, or the conduct of science is otherwise flawed, such a "weak" science might seem more susceptible to further external impositions. Conversely, science that has been conducted freely and has traditionally been held in high regard might seem more resistant. Yet this was not the case in the societies under review. In each instance scientific accomplishments had been highly respected, and the quality of science was at least as advanced as in any other country of the time. Only when the political or religious authorities pronounced immutable scientific truth did the sciences begin to suffer. An overview of the condition of science in each society bears this out.

Science in Renaissance Italy

While recognizing the importance of the seventeenth century as a scientific watershed, no date can satisfactorily mark the beginning of modern science. Empiricism became more generalized after 1600, but experimental techniques were rooted in previous centuries. Although the Renaissance was fading about the time that modern science emerged, the periods overlapped. The individualism and creativity that had characterized the Renaissance now helped to seed the scientific revolution.

During the fifteenth and sixteenth centuries in virtually every scientific discipline, Italian investigators stood at the peak. Toward the end of the fifteenth century, Leonardo da Vinci introduced techniques of observation and experimentation in the fields of anatomy, motion, and engineering that surpassed the work of his contemporaries, and his use of empirical data and systematic observation "heralded the future development of a genuine scientific method."[1]

The most important scientific works published in the middle of the sixteenth century—by Copernicus, Vesalius, and Tartaglia—were derived in large part from their authors' associations with Italian universities and Italian scholars. Though Copernicus was a Polish cleric, his studies at Bologna, Padua, and Ferrara led him to associate with Italian scientists, including Domenico Maria and Girolamo Fracastoro, whose cooperative investigations helped him to develop his interpretation of the movement of heavenly bodies.[2] Copernicus's scientific contribution was first publicly acknowledged in print in Italy. In 1561, nearly twenty years after Copernicus's death, Girolamo Ruscelli listed him for the first time as "amongst the most important astronomical observers of the time."[3]

Similarly Vesalius, a Belgian, did most of his research as a professor at Padua. Leonardo da Vinci a few decades earlier had sketched anatomical features based on dissections that he had performed and drew extremely accurate pictures of the human body, its organs, vessels, muscles, and tendons. Vesalius went further and illustrated and discussed the body in motion. His book, published in 1543, represented the first major departure from the classical descriptions of human anatomy by Galen and portrayed the interrelationship between skeleton and muscle, the body in action rather than a static depiction of its parts. For this Vesalius was recognized as "the pioneer of modern anatomy."[4]

Tartaglia's major work in kinematics, published in 1546, markedly advanced the science of moving bodies. He introduced the concept that the path of a missile was curved throughout its flight. His approach, dependent on experiment and observation rather than conjecture, was embraced by Galileo, who built on Tartaglia's propositions a half century later.[5]

In virtually every scientific discipline, Italian investigators enjoyed distinguished reputations: Cardan and Benedetti in mechanics and engineering;[6] Ghini in botany;[7] Fabrizio, Fallopio, and Eustachi in anatomy and physiology.[8] "In every department of medicine, as it was then conceived," wrote Wightman, "Padua was pre-eminent; no such lineage of teachers can be equalled elsewhere."[9]

By the end of the sixteenth century, notable scientific work was underway outside of Italy. Bacon, Gilbert, and Kepler had made seminal contributions to the experimental method, to understanding magnetism, and to planetary motion. Moreover, rating the quality of science in one community relative to another is often an elusive exercise. But, as Galileo entered adulthood, the level of scholarship and scientific inquiry in Italy was comparable with that anywhere. At the turn of the century, neither political nor ecclesiastical constraints had yet substantially interfered with scientific investigation. The élan of the earlier period of the Renaissance had dimmed, but the essential values it had fostered remained.

Early Soviet Science

During the nineteenth century Russian science was less advanced than that in other European countries. Contributions like Lobachevsky's non-Euclidean geometry, Mendeleev's periodic table, and Pavlov's work on conditioned reflex were infrequent compared to those taking place in England, Germany, and France. By the early part of the twentieth century, however, the community of Russian scientists was growing, and in the 1920s scientific expression in the Soviet Union was as unfettered as in any country of the world.

Scientists in the new Soviet state were a favored group whose projects and facilities enjoyed substantial government support. With the exception of some of the biological sciences, the quality of scientific work through World War II remained largely unaffected by ideological considerations. The sciences stood apart from other areas of Soviet life, which were subjected to large-scale interference. In consequence, "the best minds," as Graham put it, "went into scientific subjects."[10] In the 1920s, according to Zhores Medvedev, Russian science entered its "golden years."[11] During that period Einstein's theory of relativity was better received by Soviet physicists than by those elsewhere in the world, the Soviet Union became one of the world's outstanding centers for biological research, and in virtually every field of science and technology, advances were noteworthy.[12]

When the Soviet Academy of Sciences and its affiliated institutions came under control of party loyalists in the 1930s, scientists were urged to engage in research that would meet the practical needs of the state. But in most scientific disciplines, work continued with minimal state interference. Only in the late forties and early fifties was science in general threatened by state-imposed dogma.

In the years up to the official endorsement of Lysenkoism in 1948, Soviet advances in biology continued to be impressive. Following a visit to the Soviet Union in 1945, Julian Huxley summarized extensively the biological work in progress. An accomplished British geneticist, Huxley was exuberant about what he found. Despite Lysenko's influence, geneticists "continue to turn out large quantities of excellent work." He examined dozens of projects and concluded that "the U.S.S.R. is taking its place as one of the foremost countries in biological research, and I anticipate that they will soon be leading the world in [the study of] genetics and evolution."[13] Although Huxley's optimism about the immediate future was misplaced, he was an expert observer who understood what he witnessed at the time.

German Science before Hitler

The story of German science carries similar overtones. Before Hitler attained power in 1933, German science was regarded as exemplary for the rest of the world. In virtually every scientific field, German scientists were at the forefront. Albert Einstein, Max Planck, Fritz Haber, and Otto Meyerhof were world-renowned physical or biological investigators.

The medical community, later debased because of its silence during the infamous medical experiments, had been judged among the best in the world. Germany had provided many of the most illustrious names in medical research and German medicine had been "honored throughout the world."[14] German medical science was rooted in the nineteenth century traditions of Robert Koch, Emil von Behring, Paul Ehrlich, and August Bier, and was responsible for the discovery of medicines such as atabrine and serums for diphtheria and tetanus.[15] Not only individual scientists but organized medicine had developed traditions of high standards and open inquiry. By the time Hitler attained power, German medicine, like other scientific disciplines in Germany, "had served for decades as models for other communities of science."[16]

The stature of German science was demonstrated by the number of Nobel prizes won by Germans between 1901, when the first prizes were awarded, and 1932, the year before Hitler took power. Of the thirty-one Nobel prizes awarded in chemistry, fourteen had been won by Germans, five by British, four by French, and two by Americans. German scientists had received eleven of the thirty-seven physics prizes, while the British were awarded seven, the French six, and the Americans four. Six of the thirty-two prizes in physiology or medicine went to Germans, five to British, four to French, and two to Americans. In every major field, German science was distinguished.

In all the societies in the historic cases, the level of scientific accomplishment was regarded as preeminent or as advanced as in any other nation. This was true especially of the disciplines that became perverted by political authority: astronomy and mathematics in Galileo's Italy; biological sciences in Russia and Germany. Political authority imposed truths where science had previously been held in high regard, where the best scientific minds had flourished.

Science in the United States

Contemporary American science is similarly respected. During the past forty years, the United States has become the world's major center of scientific advancement. Spurred at first by an influx of European scientists who fled Nazi persecution in the 1930s, and by generous government support since World War II, American science has continued to flourish. Its supremacy is illustrated by the percentage of Nobel prizes in science awarded to Americans—54 percent between 1943 and 1981, compared to 14 percent between 1901 and 1940.

The principal difference from the earlier cases is quantitative. In the United States the number of scientists and scientific organizations and the opportunities for scientific expression are much greater. The two thousand national scientific organizations probably exceed the number of all scientists in Galileo's time.[17]

Despite growth and complexity, the organization of science in the United States shares fundamental features with that in the earlier cases. The major springboards for scientific work today are universities, government, private foundations, and industry.[18] With some qualification, this was true during Galileo's life as well as in the Soviet Union and Germany. Galileo's early activity was conducted in Italian universities, and later the Medici family integrated governmental authority with private patrimony in their sponsorship of Galileo's work. In Germany and the Soviet Union, the universities were principal centers for scientific accomplishments and private industries supported research and development (though, under Stalin, the remnants of private enterprise in the Soviet Union were eliminated).

While the growth and influence of science in the United States accelerated after World War II, the perceptions of what science is and what scientists do have changed little since Galileo's time. The experimental method—empiricism, observation, and replication—remains the principal tool in seeking scientific truth, and information about scientific activity is still disseminated by publications, letters, and at meetings largely under the auspices of scientific organizations. Indeed the most distinguished American scientific body, the National Academy of Sciences, explicitly recognizes its ties to the "great tradition" of the Lincean Academy to which Galileo belonged.[19]

The largest general scientific organization, the American Association for the Advancement of Science, is a federation of other organizations, but also claims more than 130,000 individual members. Its interests parallel those of scientific associations in Galileo's time as well as in the Soviet Union and Germany during the twenties. The objectives of the AAAS as expressed in its constitution could have been inserted in the constitutions of scientific associations in any of the earlier societies: "to further the work of scientists, to facilitate cooperation among them, to foster scientific freedom and responsibility, to improve the effectiveness of science in the promotion of human welfare, and to increase public understanding and appreciation of the importance and promise of the methods of science in human progress."[20]

Thus, apart from the difference in size, science in the United States has much in common with science in the societies of the historic cases prior to their impositions of scientific truth. As in those cases, science is at a high level of advancement. It is conducted under the auspices of government, universities, wealthy sponsors, and industry, and it is substantially subsidized by governmental funds even when conducted under nongovernmental sponsorship. Its organization and function are devoted to furthering scientific understanding, the dissemination of information, and the enhancement of communication among scientists. Its standards of investigation, experimentation, and observation remain the core of the ethos of modern science, as they were in the societies of the historic cases.

As is evident from the historic cases, neither the level of science nor the quality of scientists may be expected to act as deterrents to imposed truth. The character of a political system stands as the principal determinant. A chart (Table 5.1) at the end of this chapter summarizes conditions in the historic cases and in the United States, and makes clear the distinguishing features of their political systems. Of course no two societies are entirely alike. Industrial societies, like twentieth-century Russia, Germany, and the United States, by definition differ from preindustrial societies like seventeenth-century Italy. But in examining the question of externally imposed scientific truth, we find that the three historic cases share similar political characteristics, which provided conditions for imposing scientific truth, that do not exist in the United States.

The Political Systems Compared

To say that the United States is democratic and that the other three societies were authoritarian or totalitarian is to state the obvious. The depth of the difference between them and its importance to the question of imposed scientific truth may best be understood in a comparative review. The premise is that the historical, cultural, and institutional patterns in the United States, insofar as they inform the political process, are protectors. They provide layers of insulation against externally imposed scientific truth.[21]

POLITICAL HISTORY

Unlike the histories of the societies in the three historic cases, American political history has been relatively stable and cohesive. With the exception of the Civil War, the continuity of the political system has remained unchallenged. Pressures from racial conflicts, economic disparities, or other social stresses have strained the political order, but none has seriously threatened its existence. Continuity in American politics has been sustained by an orderly succession of officials in every branch of government since the founding of the republic.

The contrast with the societies in the historic cases is striking. Their political histories were marked by violence, by perennial invasions or threats of foreign intervention, by arbitrary rule, by the unaccountability of rulers to the citizenry, by insecurity and instability among the political leadership, by fear, terror, and intolerance of dissent.

American political history has not been devoid of rancor and tension. Many Americans—blacks, Indians, and other minorities—have suffered privations comparable to the worst in oppressive regimes. Nevertheless, most of the nation, most of the time, remained free of the relentless upheavals that plagued the societies in the three historic cases. During the century or two prior to the time of imposed scientific truth, each of those societies enjoyed periods of tranquility and political stability, but these periods were brief and fragile. Hardly a generation escaped upheavals wrought by internal revolt or foreign invasion.

POLITICAL STRUCTURE

The political structures of the three societies also bore fundamentally common features. A rigid hierarchy of power existed within each system. Atop the pyramid of power sat a single figure who was ostensibly endowed with infallible wisdom. Hitler, Stalin, and the pope each served the supposedly higher truths of their political or ecclesiastical kingdoms.

In the case of the Nazi and Soviet systems, the political party was the principal agency that established doctrines of political morality. The hierarchical structures within the National Socialist Party and the Communist Party paralleled those of the government in each society. Though government officials were often party members, administration of nonideological affairs was generally conducted by governmental units. Questions of an ideological nature fell under party authority.

The situation in seventeenth-century Italy was similar. As the institution empowered to interpret and enforce moral doctrine, the Catholic church performed the same function that the Communist and Nazi parties did. Day-to-day secular affairs, such as the administration of justice in cases of homicide and robbery, remained in the hands of civil authorities. In matters of blasphemy or heresy, the church became the exclusive arbiter. Subject to approval of the pope, the Inquisition acted as prosecutor, judge, and jury.

The structure of American political institutions stands in contrast to those in the historic cases. Political authority in the United States has been organized intentionally to separate and balance the loci of power. The system has remained structurally divided, with prescribed means to seek and to yield power. The electorate is enjoined to participate in the selection of office holders and choice of policies. Interest groups have sometimes secured advantages because of wealth, unfair manipulation, and outright corruption, but the system retains the facility for exposure and for opposing interests to surface and gain influence.

The political structure in the historic cases was centralized and authoritarian; in America it has been fragmented, balanced, and democratic. Unlike the earlier cases, America has shunned "infallible" leaders, single-party politics, and decisionmakers unaccountable to the public. Despite an occasional faltering on questions of human rights, no secret police or terror apparatus against political opponents has been institutionalized. The political structure in America has been defined by, and has helped to define, the central value that sovereignty derives from the will of the people, not from dictation by the political leadership.

POLITICAL CULTURE

A description of the political culture in any of the three societies of the historic cases is applicable to the other two as well. Political values and beliefs were rooted in authoritarian relationships in which the role of the leader was to command, and that of the follower to obey. This role relationship was both a reflection and a determinant of the values and myths that described the citizens' interactions with their political systems. When Frederick Barghoorn wrote of the Soviet leadership's efforts to "inculcate homogeneity of beliefs, perspectives, and symbols among the citizenry," he could as well have been describing the Nazi leadership or that of the seventeenth-century church.[22]

The political cultures that characterized the three societies were subject-oriented.[23] Citizens were expected to defer to their political leadership, to accept the prevailing ideology unquestioningly, and to participate politically within narrowly prescribed limits. In Italy they were expected to defer to the wisdom of secular and religious authorities. Challenging ideological or religious tenets could invite punishment or death.

Ecclesiastical authorities in seventeenth-century Italy were no more receptive to challenges of scripture than were the Soviets or Nazis to their ideological tenets. The words of an eyewitness to Galileo's condition during his trial brings to mind descriptions of the victims of the Soviets and Nazis. "It is a fearful thing to have to do with the Inquisition. The poor man has come back more dead than alive."[24]

In contrast, American political culture places a premium on broad participation in the political process. Not only voting for candidates, but the conglomeration of activities that undergird a democratic process is encouraged—dissemination of information, free press, open discussion, and the right of citizens to have access to their elected representatives.[25] While the political values in the historic cases were subject-oriented, the political culture in America is largely participant-oriented.[26]

Unquestioning political loyalty and obedience were demanded in the historic cases. American political culture holds that loyalty derives not from enforced obedience but from devotion to democratic principles. A citizen may oppose a policy or the conduct of a leader, but his loyalty to the nation is not therefore questioned, and he retains opportunities to express opposition, alter government policies, and peacefully change leaders.[27]

As violence and terror were normal instruments of rule in the historic cases, in the American political system they have been disparaged. Occasions when violence or terror have been used by American agencies to suppress dissidence have been viewed as aberrations. They are perceived as contrary to the fundamental values of American society.

IDEOLOGY

The ideologies of the societies in the historic cases governed not only the relationships of citizens to political authority but the "correct" understanding of the world and, in the case of the church, the universe. Ideology was the source of unchallengeable truth. To this end science was called forth to legitimize ideological doctrine.

Science and society suffered in Italy, not simply because there were two antagonistic versions of the truth, but in the prohibition of one version from further consideration. The ultimate authority was ideologically inspired. The tragedy was less that the church happened to choose the wrong truth, but rather that it would permit no one to "hold, defend, or teach in any way whatsoever, verbally or in writing, the said false doctrine."

Similarly in the Soviet Union and Nazi Germany, the insult to society was not merely the publicizing of warped principles. The tragedy was made manifest by the imposition of these principles in the name of science, and then prohibiting anyone from questioning them. Marx's admonition that capitalism would ultimately yield to communism "with the inexorability of a law of Nature" was not intended as metaphor.[28] The architects of Soviet society justified their repression based on this unchallengable "law of Nature." In the same manner, Nazi doctrine of racial superiority was derived from supposedly scientific evidence. "Scientific truth," wrote the Head of the Reich Bureau for Enlightenment on Population Policy and Racial Welfare, "is the basis, the justification and, at the same time, the obligation of every racial policy."[29] Despite these pretenses to the contrary, ideology dictated to science.

Here again, the American political system stands apart. To the extent that one can say an American ideology exists, it derives from the values and beliefs inherent in American political culture. Rules are not derived from revealed truths to guide political relationships. As American values and beliefs negate the existence of infallible leaders, so are they incompatible with political dogma. American political culture does incorporate ideals, including liberty and freedom of choice, that hark back to the Declaration of Independence, but unlike the historic cases, no prescribed path to fulfillment of these values has been ordained as *the* correct one.

Consequently the term "American ideology" in some respects appears contradictory. If ideology means a rigid set of beliefs and values that are translated into strict rules of political behavior, then there is no American ideology. If, on the other hand, ideology connotes a more general belief system (approximating a political culture), then American ideology places a premium on openness, broad political participation, liberty, and individual expression.

In sum, American political institutions have endured for two hundred years virtually unthreatened from within or without. The federal structure has helped promote diffusion rather than centralization of power. The ideals of democratic participation and individual worth, while not always fulfilled, remain essential values. The American political system from almost any perspective (historical, structural, cultural, ideological) is graphically different from those in the historic cases.

If this review seems to have exaggerated the virtues of the American political system, that is because of the systems it has been compared to. By some standards the American political

Table 5.1 Comparative Summary of the Three Historic Cases and the United States

POLITICAL HISTORY

Seventeenth Century Italy	Soviet Union (under Stalin)	Germany (under Hitler)	United States
Italian states racked by upheaval and violence for centuries through Galileo's time. Though Medici rule in Florence was continuous, secular and church leadership were often unstable. Political intrigue and assassinations were common.	Political violence common in Russia in 19th century. Political instability culminated in revolutions in 1905 and 1917.	Political discontent and upheaval during 19th century in Germany, though less outright violence than in Italian states or in Russia. Political instability culminated in Nazi assumption of power in 1933.	Relatively nonviolent and stable, marked by orderly succession of power.
Authoritarian political rule for centuries, shared by secular and ecclesiastical authorities.	Authoritarian political rule for centuries, by monarchy before 1917, by Communists after.	Short-lived effort to establish democratic institutions in mid-19th century; otherwise authoritarian political rule through World War I. Unstable democracy from 1919 through 1932.	Political rule based on republican forms and democratic ethos since founding of nation.
Relatively open expression permitted during 10–15 years before emergence of issues leading to imposition of scientific truth (1600–1615).	Relatively open expression permitted for ten years before emergence of issues leading to imposition of scientific truth (1922–1932).	Open expression during 10–15 years before events leading to imposition of scientific truth (1918–1933).	Generally open expression since founding of nation, though some groups, like blacks and Indians, severely circumscribed at different times and places.
Long history of productive intellectual and cultural elite. Centuries of contributions in art, writing, science, music—though restricted political expression.	Century-long history of contributions by an intellectual and cultural elite in writing, music, science—though restricted political expression.	More than a century of contributions by an intellectual elite in writing, music, science—though restricted political expression.	Opportunities for intellectual and cultural contributions generally unrestricted.

POLITICAL STRUCTURE

Seventeenth Century Italy	Soviet Union (under Stalin)	Germany (under Hitler)	United States
Authoritarian	Authoritarian	Authoritarian	Representative/ democratic
Single leader deemed infallible in spiritual matters and many temporal affairs (the pope).	Single leader infallible in all matters (Stalin).	Single leader infallible in all matters (Hitler).	Checks and balances; federated; no single source of received wisdom.
Complementary relationship between secular and ecclesiastical authorities; but with respect to the scientific truth, political judgment made by pope and Inquisition.	Political decisions imposed by narrow leadership, specifically with respect to the scientific truth.	Political decisions imposed by narrow leadership, specifically with respect to the scientific truth.	Political decisions issued by separate executive, legislative, and judicial bodies, at national, state, and local levels—generally arrived at by compromise and accommodation.
In effect the church was the dominant political "party"—the principal organ through which spiritual and much temporal power was exercised. It was the chief vehicle of political recruitment and expression of value system.	Single party rule—the Communists; chief organ for political control, propaganda, and political recruitment.	Single party rule—the Nazis; chief organ for political control, propaganda,and political recruitment.	Political parties propose programs—implementation if party representatives are elected to power. Tradition of tolerance for political opposition.
Secret police in the form of the Inquisition—spied, encouraged informers, interrogated, intimidated.	Secret police—spied, intimidated, encouraged informers, interrogated.	Secret police—spied, intimidated, encouraged informers, interrogated.	Despite occasional excesses by investigative authorities, no tradition of secret police or officially condoned intimidation.
Terror implemented against opponents and as threat to potential dissidents.	Terror implemented against opponents and as threat to potential dissidents.	Terror implemented against opponents and as threat to potential dissidents.	Terror as political instrument not generally tolerated.

POLITICAL CULTURE

Seventeenth Century Italy	Soviet Union (under Stalin)	Germany (under Hitler)	United States
Subject-oriented citizenry, though politically informed.	Subject-oriented citizenry, though politically informed.	Subject-oriented citizenry, though politically informed.	Participant-oriented citizenry—informed, often active.
Loyalty and obedience to regime required of citizenry. Open discussion discouraged in secular and ecclesiastical matters.	Loyalty and obedience to regime required of citizenry. Open discussion of political questions discouraged.	Loyalty and obedience to regime required of citizenry. Open discussion of political questions discouraged.	Loyalty to democratic ethos encouraged; usually open discussion of political questions.
Political paranoia on part of leadership. The church feared the impact of the Reformation and increasing secular power; the Jesuits feared losing control of scholarship and science within the church.	Political paranoia on part of leadership. Before 1917 several political parties opposed the Bolsheviks. After their suppression, factions contended through the 20s, and latent opposition frightened Stalin through the 30s and 40s resulting in party purges.	Political paranoia on part of leadership. Nazis faced competition from other parties before 1933. Afterward, intraparty factionalism and jealousies prompted purges by leadership.	Regular, institutionalized contests for leadership. Openness and exposure of candidates' beliefs and attitudes encouraged.
Political violence was common and justified as a legitimate instrument of rule.	Political violence was common and justified as a legitimate instrument of rule.	Political violence was common and justified as a legitimate instrument of rule.	Political violence has not been common, and has not been considered a legitimate instrument of rule.

IDEOLOGY

Seventeenth Century Italy	Soviet Union (under Stalin)	Germany (under Hitler)	United States
Ideology cited as highest authority	Ideology cited as highest authority.	Ideology cited as highest authority.	No official ideology.
Ideology of the church held that the individual counted for little. Private interests were to be sacrificed to the church's.	Soviet ideology demanded the sacrifice of individual interests to the state. Contrary to Marx, all citizens were made subjects of an increasingly powerful state.	Nazi ideology demanded absolute obedience of citizenry to the state. The interests of the state were to be determined by the political leadership exclusively.	Though imperfectly applied, responsiveness of leaders to the will of the people is central value.
The sacrifice of individual interests to the church's, promised reward in heaven for those who obeyed.	The sacrifice of individual interests to the state's, meant a better world for all (later generations).	The sacrifice of individual interests to the state's, meant a better world for the privileged as defined by Nazi racial doctrine.	Belief in the inherent worth of the individual as an ideal.
"Inferior" people were defined ideologically by religion. Non-Christians in general were perceived as inferior, nonmembers of the church in particular, though individuals could change categories by embracing the church.	"Inferior" people were defined ideologically by class. Capitalists and bourgeoisie were undesirables. At first some were tolerated if their skills were helpful to the regime. In the 20s and 30s many were killed in purges.	"Inferior" people were defined ideologically by race. Aryans were explicitly deemed superior to all others, while Jews and gypsies were to be exterminated. Racial categories were immutable—individuals could not change.	No people or groups defined officially as inferior. Though treatment of some groups has been discriminatory, the national ideal calls for equal treatment of all people.

SCIENCE QUALITY AND CHARACTER

Seventeenth Century Italy	Soviet Union (under Stalin)	Germany (under Hitler)	United States
Beginnings of era of modern science. Principles of experimentation, observation, and reason over faith and intuition were understood and employed.	Principles of modern science had long been understood and employed.	Principles of modern science had long been understood and employed.	Principles of modern science long understood and employed.
Quality of science in general considered excellent.	Quality of science in general considered excellent.	Quality of science in general considered excellent.	Quality of science in general considered excellent.
Quality of physics and astronomy, the subjects of the imposed truth, was unsurpassed in the world prior to politically dictated truth.	Quality of genetics and plant physiology was unsurpassed in the world prior to politically imposed truth.	Quality of medicine in all aspects—application, teaching, research—was unsurpassed in the world prior to politically imposed truth.	Quality of almost all disciplines involved in the historic cases—physics, astronomy, genetics, medicine—unsurpassed in the world.
During 10–15 years before beginning of events leading to the imposition, scientific debate was unrestricted, and dissemination of scientific information was broad and uninhibited. (Until 1616.)	During 10–15 years before movement toward imposition of truth began, scientific debate was unrestricted and dissemination of scientific information was broad and uninhibited. (During 20s and early 30s.)	During 10–15 years before beginnings of imposed scientific truth, scientific debate was unrestricted and dissemination of scientific information was broad and uninhibited. (During 20s and early 30s.)	Scientific debate traditionally unrestricted and largely uninhibited.
The imposed truth did not immediately affect other areas of science, especially biological and medical research, which continued to flourish. But after a few decades scientific activity shrank in general, and Italy never regained primacy.	During time of imposed truth other sciences continued to develop, though occasionally burdened by state dictation. In 1948–1950 several disciplines were pressed to conform to "truths" imposed by party spokesmen, but otherwise science developed with minimal outside interference.	Some science continued to develop, though almost all disciplines suffered from dismissal of many scientists for racial or political reasons.	No threat of politically imposed scientific truth.

SCIENTIFIC ORGANIZATION

Seventeenth Century Italy	Soviet Union (under Stalin)	Germany (under Hitler)	United States
The Lincean Academy in early 17th century was the first modern scientific society. It functioned much as its successors have: held meetings, encouraged exchange of scientific information, sponsored publications, provided forums for differing points of view. Political discussion, however, was discouraged.	Various scientific disciplines were organized in separate societies, though all under auspices of Soviet Academy of Sciences. In 1930s and 40s, discussion of scientific matters permitted if not contrary to political or ideological dictate.	In 1930s, efforts by regime to monitor science varied from trying to control traditional societies to forming new compliant ones. Scientific debate continued, however, though opposition to party line was not permitted.	More than 2,000 national scientific organizations plus many regional and local groups that promote dissemination of information, discussion, and debate. Sometimes asked for scientific advice by political leaders.
Pressure was brought by church to conform to political line and to oppose Copernicanism. Ultimately the Academy disbanded.	Pressure brought by regime on scientific academies to conform to political line. The academies yielded.	Pressure brought by regime on scientific societies to conform to political line. The societies yielded.	Sometimes subjected to pressures from outside political sources, but never sustained or national effort to adopt externally mandated truths.

THE POLITICALLY IMPOSED SCIENTIFIC TRUTHS

Seventeenth Century Italy	Soviet Union (under Stalin)	Germany (under Hitler)	United States
The imposed scientific truth was not suddenly introduced but was rooted in years of earlier decisions. (Copernicus; Bruno)	The imposed scientific truth was not suddenly introduced but was rooted in years of earlier decisions. (Michurin)	The imposed scientific truth was not suddenly introduced but was rooted in years of earlier decisions. (19th century racist philosophers)	No threat of politically imposed scientific truth.
The imposed truth was supported by the dominant political organization, the church, and within the church by its strongest faction, the Jesuits.	The imposed truth was supported by the dominant political organization, the Communist party.	The imposed truth was supported by the dominant political organization, the Nazi party.	No threat of politically imposed scientific truth.
The imposed truth was justified as conforming to ideology—the sanctity of God and the Bible.	The imposed truth was justified as conforming to ideology—Marxism-Leninism, dialectical materialism.	The imposed truth was justified as conforming to ideology—racial science was a defining principle of Nazism.	No threat of politically imposed scientific truth.

THE SCIENTISTS

Seventeenth Century Italy	Soviet Union (under Stalin)	Germany (under Hitler)	United States
Most scientists were aware of the issues, had access to the facts, and knew of the church's dim view of heliocentrism during the 1610s and 1620s.	Most scientists were aware of the issues, had access to the facts, and knew of the party's warm tolerance of Lysenkoism during the 1930s and 1940s.	All scientists were aware of Nazi racial science. Thousands had access to information about the medical experiments during the 1940s.	Virtually all scientific activity is reported and available to the scientific community.
A few scientists vigorously opposed Galileo and his Copernican views; a few actively supported them. Most remained silent.	A few scientists vigorously opposed Lysenko and his views; a few actively supported them. Most remained silent.	A few scientists vigorously opposed Nazi racial science; a few actively supported it. Most remained silent.	Most scientists would not likely oppose an effort by the state to impose scientific doctrine.
If many scientists had openly supported Galileo, their pressure might have altered the church's position, according to scholars of the question. Until the trial, support for Galileo could draw criticism from the church, but no one suffered penalties beyond that.	Until the party decree in favor of Lysenko in 1948, criticism of Lysenkoism was legitimate, though it resulted in loss of position for some scientists.	Scientists who questioned the methodology of the medical experiments might have risked their positions or official favor, but probably little else.	Scientists have been reluctant to oppose forthrightly political actions less threatening than state mandated truth.

system has responded inadequately to the needs of the citizenry. Economic deprivation and discrimination persist and special interests are able to gain and wield influence at the expense of others.

Yet inferences that the American political system is comparable to those in the historic cases ignore the many profoundly important institutional differences. These differences are so palpable that spelling them out may seem gratuitous, but the exercise is warranted, for many observers mistakenly believe that American political institutions offer little more protection against imposed scientific truth than those in the regimes of the historic cases. To regard all political systems as threats to scientific activity is to ignore differences among political systems. Political institutions that loom fearsome in some societies may be the catalysts of legitimate scientific activity in others.

Notes

1. Gene A. Brucker, *Renaissance Italy* (New York: Holt, Rinehart and Winston, 1958), p. 48.
2. W.P.D. Wightman, *Science in a Renaissance Society* (London: Hutchison and Co., 1972), p. 116.
3. Peter Laven, *Renaissance Italy* (New York: G. P. Putnam's Sons, 1966), p. 185.
4. Bertrand Russell, *A History of Western Philosophy* (New York: Simon and Schuster, 1945), p. 544.
5. Laven, pp. 175–76.
6. Stephen F. Mason, *Main Currents of Scientific Thought* (New York: Abelard-Schuman, 1956), pp. 118–19.
7. W.P.D. Wightman, *Science and the Renaissance*, Vol. 1 (New York: Hafner Publishing Co., 1962), pp. 194–95.
8. Laven, p. 193; Fielding H. Garrison, *History of Medicine*, 4th ed. (Philadelphia: W. B. Saunders Co., 1963), p. 234.
9. Wightman, *Science and the Renaissance*, p. 207.
10. Loren R. Graham, *Science and Philosophy in the Soviet Union* (New York: Vintage Books, 1974), p. 432.
11. Zhores A. Medvedev, *Soviet Science* (New York: W. W. Norton and Co., 1978), p. 13.
12. *Ibid.*, p. 17; Graham, p. 196; David Joravsky, *Soviet Marxism and Natural Science, 1917–1932* (New York: Columbia University Press, 1961), pp. 275–76; David Joravsky, *The Lysenko Affair* (Cambridge, Mass.: Harvard University Press, 1970), p. 107.
13. Julian Huxley, "Science in the U.S.S.R.: Evolutionary Biology and Related Subjects," *Nature*, vol. 156 (September 1, 1945), pp. 254–56.

14. *Trials of War Criminals before the Nuernberg Military Tribunals under Control Council Law No. 10,* "The Medical Case," Vols. 1 and 2, October 1946-April 1949 (Washington, D.C.: U.S. Government Printing Office), p. 56.

15. *Ibid.,* Vol. 2, p. 90.

16. Joseph Haberer, *Politics and the Community of Science* (New York: Van Nostrand Reinhold Co., 1969), pp. 103–104.

17. 2069 national organizations categorized as scientific, engineering, technical, health, or medical are listed in *Encyclopedia of Associations, 11th ed., National Organizations of the United States,* Vol. 1 (Detroit: Gale Research Co., 1977), pp. 337–426, 711–819. The total does not include regional, state, or local societies.

18. James L. Penick, Jr., Carroll W. Pursell, Jr., Morgan B. Sherwood, Donald C. Swain, eds., *The Politics of American Science, 1939 to the Present,* rev. ed. (Cambridge, Mass.: The M.I.T. Press, 1972), pp. 6–9.

19. The National Academy of Sciences [descriptive folder] (Washington, D.C.: The National Academy of Sciences, August 1977).

20. Constitution of the American Association for the Advancement of Science, article 2, section 1, in *Handbook, 1979* (Washington, D.C.: AAAS Publication, 1979), p. 140.

21. The political characteristics ascribed to each society in this discussion are recounted in many works. They may be gleaned from publications cited throughout this chapter, as well as from other sources, including John Addington Symonds, *Renaissance in Italy* (New York: Cooper Square Publishers, Inc., 1966); J. Lucas-Dubreton, *Daily Life in Florence in the Time of the Medici,* translated by A. Lytton Sells (New York: Macmillan Co., 1961); Eric Cochrane, *Florence in the Forgotten Centuries 1527–1800* (Chicago: University of Chicago Press, 1973); Niccolo Machiavelli, *The Prince* (New York: Mentor, The New American Library, 1962); Karl Dietrich Bracher, *The German Dictatorship,* translated by Jean Steinberg (New York: Praeger Publishers, 1970); Lewis J. Edinger, *Politics in Germany* (Boston: Little, Brown and Co., 1968); William L. Shirer, *The Rise and Fall of the Third Reich* (New York: Simon and Schuster, 1960); Merle Fainsod, *How Russia Is Ruled,* rev. ed. (Cambridge, Mass.: Harvard University Press, 1963); John H. Hazard, *The Soviet System of Government,* 3d ed. (Chicago: University of Chicago Press, 1964); Leonard Schapiro, *The Communist Party of the Soviet Union* (New York: Vintage Books, 1964); Adam B. Ulam, *The Bolsheviks* (New York: The Macmillan Co., 1965).

22. Frederick C. Barghoorn, *Politics in the USSR* (Boston: Little, Brown and Co., 1966), p. 13.

23. Gabriel A. Almond and Sidney Verba have delineated three types of political culture: parochial, subject, and participant. For extended discussion see their book, *The Civic Culture* (Princeton, N.J.: Princeton University Press, 1963), esp. chap. 1.

24. Comment by the Florentine ambassador to Rome, in Giorgio de Santillana, *The Crime of Galileo* (Chicago: University of Chicago Press, 1955), p. 258.

25. A classic statement of the social bases of democracy is in Seymour Martin Lipset, *Political Man* (Garden City, N.Y.: Doubleday and Co., 1960), esp. chap. 2.

26. Almond and Verba, pp. 313–15.

27. The importance of political socialization as a determinant of political values and behavior is stressed in *ibid.* The process in America is reviewed in James David Barber, *Citizen Politics* (Chicago: Markham Publishing Co., 1969).

28. Karl Marx, "The Revolutionary Transformation of Capitalism," in C. Wright Mills, *The Marxists* (New York: Dell Publishing Co., 1962), p. 69.

29. Statement by Dr. Walter Gross, Head of the Reich Bureau for Enlightenment on Population Policy and Racial Welfare, in *Germany Speaks* (London: Thornton Butterworth Ltd., 1938), p. 73.

PART THREE

The American Political System and the Restraint of Science

6
Imposed Truth

The American political system, as has been emphasized, differs markedly from the political systems of seventeenth-century Italy, Stalin's Russia, and Hitler's Germany. This chapter will highlight the significance of this difference in dealing with attempted impositions of scientific truth in this country. The attitudes of American scientists about the role of a political system in the perversion of science are discussed, based on their responses to a questionnaire that sought their perceptions about a variety of issues. The methodology and description of the sample are included in the Appendix.

Contrary to the views expressed by most respondents, this book argues that American institutions are protectors against impositions comparable to those in the historic cases. Nascent attempts to impose scientific truth here, such as efforts to mandate the correctness of the Bible's explanation of man's origin and to pass "human life" legislation, bore features reminiscent of the historic cases. Their relationship to American political institutions is reviewed to show how the political system has acted to protect society from imposed scientific truth.

Importance of Political Systems

The importance of a political system relative to the imposition of a scientific truth is appreciated by the scientists in this survey. In Table 6.1, only 18 percent feel that science or scientists are principal determinants. Seventy-two percent believe that a society's political system, including the values and

Table 6.1 Responses of Scientists to Question 25 of the Survey (by percent)

The principal determinant of whether a scientific perversion might be imposed in any society lies in the nature of:

the Scientific Discipline in Question	3
Scientists in that Society	10
The Organization of Scientists in that Society	5
The Political System*	72
Other/Can't Say/No Answer	10
Total	100 (632)

*Included in this category are 4 percent who offered multiple answers, one of which was the political system, and 5 percent who wrote in "values and beliefs of the citizenry."

Note: In this and subsequent tables the percentages are rounded off and totals may not be 100.

Table 6.2 Responses of Scientists to Question 14 of the Survey (by Percent)

Historic perversions of science like Nazi racial science, Soviet biology under Lysenko, and unchallengeable dogma that the earth was the center of the universe would be virtually impossible to impose in American society.

Strongly agree	Agree	Disagree	Strongly disagree	Can't say/ no answer	Total
7	30	47	10	7	101 (632)

beliefs of its citizens, stands as the principal determinant of whether science is likely to be perverted or not.

Could imposed truths like those in the historic cases be inflicted on American society? As alluded to in the introductory chapter, 57 percent of the respondents in Table 6.2 think it possible that perversions like Nazi racial science, Lysenkoite biology, or dogma that the earth is the center of the universe could be imposed here. Only 37 percent reject the possibility.

The view of the majority parallels that of scientists quoted earlier: that restraint of scientific activity in any political system threatens to pervert all scientific activity. Respondents who elaborated on the point seem skeptical not about the ideals of the American system, but about current values and the system's ability to live up to its ideals.

One respondent doubts that the American public is now sufficiently "well-informed and educated" to resist political or religious dogma. A second believes that the "faddishness of U.S. culture and society" makes it susceptible. Another worries that "an ever-increasing anti-intellectual atmosphere combined with severe economic pressures" render us vulnerable. A fourth believes that the "imposition of a scientific perversion can be prevented only by a knowledgeable and free press. But we do not have that in the U.S. today—free, yes; knowledgeable, no."

Several scientists offered lengthy comments prompted by these questions. Although mistrustful of the American political system, their reasons often appeared contradictory. A university physicist, for example, deplored the system for spawning harmful activities carried out in the name of science, implying that the system might be prone to impose scientific truths as well. He said:

> It is of course true that in the U.S. the control of ideas generally, and of scientific work in particular, is not as nearly total as these controls can be in a tightly-run dictatorship. There is more room for dissent here. But that ought not allow one to feel comfortable, for this society is almost overwhelmingly dominated by very powerful forces—constellations of people and wealth—which are committed to the ideology of exploitation, and this ideology permeates the communications media to such an extent as to be practically unchallenged. Possibly the most dangerous "perversion of science" is its use to legitimize the calculus of intervention and manipulation on a massive scale: Diversion of rivers, climate modification, changes of atmospheric composition, genetic engineering, dissemination of radioactive materials in the biosphere. Science could serve human, and humane needs, but it would need to be science as practiced in a humane society. . . .

While this scientist scorned abuses by private interests and the permissive system that encourages them, others were skeptical about the government for opposite reasons. An industrial chemist deplored what he considered unwarranted governmental intrusions that are "frequently disastrous rather than beneficial." He enclosed a letter in defense of his company's products, in which he writes:

> Most of the concern about lead in paints goes back from 50 to 100 years during the period when white lead was an important pigment in making both interior and exterior house paints. . . . This type of paint became obsolete in the early 20s when titanium dioxide became commercially available with approximately 6 times the hiding power per pound, no tendency at all to form a black compound in the presence of sulphur, and completely non-toxic when ingested.
>
> In spite of this well established history of coatings technology, our politicians are approximately 100 years behind the facts and have done nothing but perpetrate consumer fraud with the present lead scare and legislation.

Both scientists reveal keen dissatisfaction with the United States government's role regarding scientific and technological activities. Yet each is critical for contrary reasons—one because the government is too permissive with private industry, the other because it is too restrictive—and their comments represent the confusion that many scientists expressed about political institutions. But if the political bearings of many scientists were in conflict, their feelings about scientific truth were nearly unanimous.

Imposed Truth: A Perversion of Science

Since the publication of Thomas Kuhn's *The Structure of Scientific Revolutions* in the early 1960s, philosophers and historians increasingly have argued that scientific truth is inextricably tied to language, culture, and values.[1] Scientific truth cannot, according to this view, be understood as objective. As may be seen from Table 6.3, responding scientists overwhelmingly reject this proposition. Eighty-eight percent believe (56 percent strongly) that scientific truth exists apart from human values, and 85 percent hold that science, more than other disciplines, is structured on objective truths.

The definition of a perversion of science as used in this study is broadly accepted as valid. Ninety-five percent in Table

Table 6.3 Responses of Scientists to Questions 7 and 12 of the Survey (by Percent)

	7. Genuine scientific truth exists apart from human values.	12. Science more than other disciplines is structured on objective truths
Strongly agree	56	26
Agree	32	59
Disagree	5	8
Strongly disagree	4	1
Can't say/ no answer	4	1
Total	101 (632)	100 (632)

Table 6.4 Responses of Scientists to Question 8 of the Survey (by Percent)

The imposition of scientific truth by a political system and the prohibition of challenges to that truth may aptly be characterized as a perversion of science.

Strongly agree	Agree	Disagree	Strongly disagree	Can't say/ no answer	Total
71	24	1	1	3	100 (632)

6.4 agreed (71 percent strongly) with the definition employed here. This constitutes a larger measure of agreement than for any other response item, and supports a defining principle of this study—that politically imposed scientific truth constitutes a perversion of science.

Thus, despite the debate among some scholars about the inability to separate scientific truth from human values, the issue appears to hold little ambiguity for most American scientists. Even more clear-cut is their agreement about what constitutes a perversion of science.

Table 6.5 Responses of Scientists to Question 9 of the Survey (by Percent)

A scientist's training and experience provide an advantage over nonscientists in ability to analyze issues.

Strongly agree	Agree	Disagree	Strongly disagree	Can't say/ no answer	Total
13	48	22	4	14	101 (632)

Table 6.6 Responses of Scientists to Question 10 of the Survey (by Percent)

Scientists are more likely than nonscientists to reject unsubstantiated claims of truth or fact.

Strongly agree	Agree	Disagree	Strongly disagree	Can't say/ no answer	Total
19	59	13	3	6	100 (632)

If, as the scientists believe, scientific truth and its perversion are knowable, the responses also make it clear that scientists think they are in the best position to recognize them. In Table 6.5, 61 percent believe they are better equipped than nonscientists by virtue of training and experience to analyze issues in general, and in Table 6.6, 78 percent think they are more likely than nonscientists to reject unsubstantiated claims of truth or fact. If their perceptions are accurate, scientists should be the first to become suspicious about efforts to tamper with the truth. They should be the first to recognize the threat of politically imposed scientific truth.

How would American scientists behave in the face of such an imposition? Would they, as Robert Oppenheimer believed, "yield a little more slowly than others to the natural corruptions of their time"?[2]

In turning to a review of nascent impositions of scientific truth in American society, we find that, contrary to Oppenheimer's supposition, American scientists have acted little differently from scientists in the historic cases. Protection of the citizenry ultimately has come from the structures, values, and beliefs that comprise the American political system.

Antievolution

> Be it enacted by the General Assembly of the State of Tennessee, That it shall be unlawful for any teacher in any of the Universities, Normals and all other public schools of the State which are supported in whole or in part by the public school funds of the State, to teach any theory that denies the story of the Divine Creation of man as taught in the Bible, and to teach instead that man has descended from a lower order of animals. (Chapter 27, House Bill 185, Public Acts of Tennessee for 1925)

House Bill 185 was passed in the Tennessee House by a vote of 71 to 5, in the Senate 24 to 6, and was signed into law by Governor Austin Peay on March 21, 1925. Four months later John Thomas Scopes sat in a sweltering courtroom in Dayton, Tennessee, accused of having violated the statute.[3]

Scopes had assigned readings on Darwin's theory of evolution to his high school biology class. Now his trial was being celebrated as a grand confrontation between science and religion, between knowledge and ignorance. The principal protagonists, Clarence Darrow for the defense, William Jennings Bryan for the State, embraced the issue on these terms, though they differed about who was possessed of knowledge and who was guilty of ignorance.

The effort to prohibit the teaching of evolution shared several characteristics with the historic cases. Political and religious authorities sought to mandate scientific truth in Tennessee as they did in the other societies, but while the law remained on the books for forty-two years, it remained only a latent imposition that did not materialize.

BACKGROUND

Tennessee was one of several states immersed in the evolution issue during the twenties. Fundamentalist religious conviction held that the Bible was a literal chronicle of history, and this belief was embedded in much of the South as well as other rural parts of the nation. In addition to Tennessee, Mississippi enacted an antievolution law in 1927, and in 1928, after the Arkansas state legislature failed to enact a similar law, the citizens of that state voted to adopt one by a margin of two to one.[4] Between 1921 and 1929 fundamentalists introduced thirty-seven antievolution bills in twenty state legislatures, though almost all were defeated.[5]

Antievolution never became national dogma, nor, except in a few states, did it receive the sanction of law. On that basis alone it is not equivalent to the impositions of truth that occurred in Germany, Russia, or seventeenth-century Italy. While the Tennessee law forbade teaching evolution in state-supported schools, the proscription stopped there. It was neither as pervasive nor as absolute as the historic cases. Clarence Darrow recognized the greater threat beyond the existing law in a statement he made during the trial:

> If today you can take a thing like evolution and make it a crime to teach it in the public school, tomorrow you can make it a crime to teach it in the private schools, and the next year you can make it a crime to teach it to the hustings or in the church. At the next session you may ban books and newspapers. Soon you may set Catholic against Protestant and Protestant against Protestant, and try to foist your own religion upon the minds of men. If you can do one, you can do the other. Today it is the public school teachers, tomorrow the private. The next day the preachers and lecturers, the magazines, the books, the newspapers.[6]

While the antievolution issue was of smaller dimension than the historic cases, it bore sufficient similarities to test the American political system in the face of an effort to impose scientific truth.

THE POLITICAL SYSTEM AND EVOLUTION

Scopes was convicted and fined $100.00. Two years later, while holding the act constitutional, the Tennessee Supreme Court reversed the conviction on a technicality about the manner in which the fine was levied. The court recommended that Scopes not be retried, to preserve the "peace and dignity of the state."[7] The act was finally repealed in 1967.

Despite retention of Tennessee's and a few other state laws for more than four decades, none was effectively implemented.

The common view, offered retrospectively by a scientist who had worked with the defense, was that "Tennessee had been made to appear so ridiculous in the eyes of the nation that other states did not care to follow its lead."[8] Others have affirmed that the cause of evolution was not really advanced, that in Tennessee evolution was hardly taught after Scopes.[9] Moreover, between 1925 and 1960, discussion of evolution declined in high school biology textbooks used throughout the nation.[10]

Although the antievolution movement did not die at Dayton, an overview of the fifty-five years since the Scopes trial points to a creditable response by the American political system. While some state laws were retained, none was enforced. Even in Tennessee, where teaching evolution might draw the wrath of local fundamentalist preachers, an enterprising teacher could find ways around the problem.[11] Pockets of the nation may have remained inhospitable to Darwinism, but no one after the Scopes trial was arrested, charged, or tried for teaching Darwin's theories.[12]

The failure of the effort to impose fundamentalism as public doctrine is a favorable comment on the political system. The fact that the issue has not "finally" been resolved does not weaken the assertion. Creationists continue to make the fundamentalist case, as we shall discuss, but it is the strength of the system that has allowed them to continue while denying them the right to impose their doctrine.

The fact that the fundamentalist effort has not gone beyond its nascent character to full legitimation is politically instructive. First, the issue has been openly discussed since its origin. The citizenry has been thoroughly exposed to the arguments, sometimes tiresomely.[13] Second, all sides have been heard in courts of law. While the impartiality of the court in the Scopes trial was in question, the effort of exposure ultimately rendered a just response. Third, despite the passage of antievolution laws in a few states, no governmental authority at any level successfully imposed scientific truth on the citizenry. Finally, while in some schools evolution may not be part of the formal curriculum, issues on all sides can be discussed in or out of schools without legal restraint.

The evolution issue has drawn on every feature of the political system. American political structures, habits, values, and beliefs all joined to affect the outcome. The influence of structure was most evident in the rejection by almost all state legislatures of any antievolution laws. It was also central to the Tennessee Supreme Court's ruling on the Scopes case. An appeal beyond Tennessee to the United States Supreme Court would likely have led to a finding that the antievolution law was unconstitutional, something the Tennessee court wished to avoid.[14] Accordingly, the Tennessee Supreme Court not only reversed Scopes's conviction on a technicality, but it also advised against retrial. The court preferred an emasculated law to abrogation. Nevertheless the effect was the same.

Insofar as American political history and habits reflected on the debate, while the issue was very divisive, it prompted no terror, no violence. The arguments were presented openly. In the long run the disposition of the issue reflected values of the general citizenry. No perversion of science took place because the American citizenry could express its will through its political and social institutions. The institutions that encouraged open inquiry and reasoned deliberation, however imperfectly at some stages, ultimately proved successful.

THE SCIENTISTS' ROLE

As posed by the state of Tennessee in 1925, antievolution was an effort to force people to ignore convincing scientific evidence. By the turn of the century the theory of evolution had been broadly accepted among scientists.[15] There was little difference in the convictions of scientists about the issue in 1925 than there is today. Although fossil records, dating techniques, and understanding of comparative anatomy are more complete today, they were tools of evidence in the twenties as well.[16]

A great majority of respondents in our survey view antievolution laws in states as comparable to the historic cases of imposed scientific truth, and similar thinking undoubtedly prevailed among scientists during the twenties. In response to our questionnaire, 76 percent agreed that laws prohibiting the teaching of evolution constituted perversions of science. Among the 21 percent who disagreed, many made clear that their disagreement was a matter of scale, not substance. If

Table 6.7 Responses of Scientists to Question 17 of the Survey (by Percent)

The prohibition of teaching the theory of evolution in several states constituted a scientific perversion comparable to the historic cases cited in question 14 (Nazi racial science, Soviet biology under Lysenko, and unchallengeable dogma that the earth was the center of the universe).

Strongly agree	Agree	Disagree	Strongly disagree	Can't say/ no answer	Total
29	47	19	2	4	101 (632)

the issues were not comparable it was because the question was about state laws rather than a federal law (Table 6.7).

Most scientists in the twenties must have viewed the antievolution laws as threats to scientific expression and inquiry. The Scopes trial presented an unusual opportunity for scientists to rally and lend support to the defense. A few did. But the "older and more eminent" scientists had declined the opportunity, according to a historical account.[17] A scientist who worked with the defense later recalled his surprise "when I learned that several of the elder statesmen of zoology had been first solicited as witnesses by the American Civil Liberties Union but had made excuses of various sorts."[18] Some apparently worried that open support of Scopes might adversely affect their careers, despite the potentially devastating effect that the law might have on science as a whole.

Criticism of the scientific community's response to antievolution was more recently raised in another context. Grabiner and Miller found that most high school biology textbooks throughout the nation contained less information about evolution during the three decades following the Scopes trial than before it. The authors blame the scientific community for its "large-scale failure to pay attention to the teaching of science in the high schools."[19]

Even in the sixties, when textbook content about evolution had increased, the authors gave scant credit to efforts by the scientific community. Rather, interest on the part of scientists was tucked among political and social currents that the authors cited as more instrumental. The prestige, power, and financial support of the federal government were principal reasons, according to Grabiner and Miller. "In addition, major historical changes had occurred . . . : the new public interest in improving high school science teaching; the large body of legal precedents limiting religious influence in the schools; and the increasing urbanization and educational level of the people in the south. These same historical forces resulted in the repeal of the Tennessee antievolution law in 1967."[20]

Creationism

Challenges to the American political system involving politically imposed scientific truth did not end in the 1920s. During the past decade the fundamentalist persuasion was resuscitated in the guise of "creation science." Rather than simply ban the teaching of evolution as the fundamentalists urged, the creationists have sought to spread their tenets more obliquely. In two ways they approach their task differently from the fundamentalists. They hold that creation science is a valid interpretation of the origin of life, that its validity stands independently of religious belief. Creationism is scientifically true, they say, and just happens to correspond with the Bible's story of creation. By arguing in these terms, creationists seek to avoid the First Amendment stricture against teaching religion in public schools.

Additionally, creationists maintain that while their views are valid and those of the evolutionists are not, all they want is equal time in public schools to present their case. Students, they argue, will then be able to judge for themselves which of the two explanations is valid. Despite this facade, the creationists' approach seeks the political imposition of scientific truth no less than did the fundamentalists'.

Contrary to the impression that creationists try to generate, their tenets parallel those of biblical fundamentalism: God created the cosmos as an act of divine will; the earth and all existing life forms were created at about the same time; the earth is only a few thousand years old, not 4.5 billion as conventional science has determined.[21] Creationists dismiss the notion that life evolved through natural selection, and discount the validity of radioactive decay as a measure of age by maintaining that radioactive materials are only a few thousand years old.[22] They hold that, despite outward appearances, the earth is not "really old . . . just tired."[23] Fossils and bones of supposedly prehistoric man are dismissed either as frauds, or as the remains of apes or of once-normal human beings who contracted bone disease.[24]

During the 1970s, the creationists generated considerable interest in their cause. By 1980 more than twenty states were considering legislation that would mandate the teaching of creationism in public schools where evolution was taught. In 1981, Arkansas and Louisiana enacted such legislation. The laws in both states were challenged, and many thought they would see a replay of the Scopes trial. Once again the ACLU led the legal challenge. Once again state law in effect mandated what was scientifically "correct." The equal-time provision meant that creationism *had* to be taught under specified circumstances.

In other ways the trials were very different: In Arkansas there were no theatrics and the case was tried without a jury before a federal district court judge who was not predisposed toward the creationists. Judge William Overton decided in uncompromising language that the Arkansas law violated the Constitution. He ruled that scientific creationism was nothing more than the biblical version of Genesis, and that to teach it in public schools would violate the First Amendment clause against the establishment of religion. The verdict was so patently dismissive of the creationists' position that the state did not appeal. Creationists blamed the defeat on a poorly prepared case by the state attorney general and on a biased judge. They vowed to win in Louisiana.[25]

SCIENTIFIC AND POLITICAL RESPONSE

As happened before, the reaction of scientists to this challenge to scientific truth was sluggish and apathetic. A small group of science teachers formed "committees of correspondence," and warned that religious doctrine could replace scientific theory in public school classes. Few scientists heeded.

At the end of 1981, after enactment of the Arkansas and Louisiana bills, and the introduction of similar bills in twenty other states, a concerned scientist despaired that "this is an extremely important issue, but so far it has been ignored by much of the scientific community."[26] As happened fifty-five years earlier at the time of the Scopes trial, a few scientists spoke out and some organizations issued appropriate statements. But most scientists went about their affairs oblivious to warnings that "it's not just biology that's in danger, it's all of science: geology, physics, astronomy. The creationists are attempting to mandate what is appropriate for study and what is not."[27]

A handful of scientists testified for the plaintiffs at the Arkansas trial, but once again the principal protector of the general citizenry from politically imposed scientific truth proved to be the network of institutions that comprises the American political system. In the Arkansas case the judge ultimately rested his opinion on the authority of the Constitution and on the Supreme Court's previous decisions regarding the teaching of evolution.[28] In a broader sense, the public was protected by a series of institutional safeguards against popular whim or personal beliefs of a transient political leadership.

President Reagan had made clear that he supported the creationist movement. During the 1980 presidential campaign he said, erroneously, that scientists were increasingly skeptical about the validity of evolution and that "the biblical story of creation should also be taught."[29] Many newly elected conservatives in Congress thought similarly. Most people who were surveyed on the subject agreed that creationism should be taught along with evolution in public schools.[30] But underlying the American belief system is a tacit understanding that great principles, such as scientific truths, are not mandated by politicians or opinion polls. Insofar as the Arkansas case involved constitutional issues, Judge Overton rejected appeals to political whimsey. "The application and content of First Amendment principles are not determined by public opinion polls or by a majority vote," the judge wrote. Whether proponents of creationism "constitute the majority or minority is quite irrelevant."[31]

The creationists' frustration with what they see as institutional obstacles was manifested by a staff member of the Institute for Creation Research during an interview. Why not seek congressional legislation for creationism rather than face painstaking efforts state by state? The response: "Because states are in charge of educational policies, not the federal government, and rules in one state are different from another."[32] The federal structure may loom as an obstacle to creationists, but it stands as a safeguard to the overall population.

A word is appropriate about the creationists' call for fairness. They claim they wish only to expose their views, not to dictate them. The argument is invalid, however, for even apart from the constitutional issue regarding separation of church and state, the best scientific evidence at a particular time self-evidently deserves primacy in a curriculum.

In sum, a few scientists provided important assistance at the Scopes and creationist trials, and some issued public statements of support. But whether in the 1920s or in subsequent decades, scientists have been reluctant to join in battle against the threat of a politically imposed scientific truth. Most kept a safe, silent distance. Their behavior was like that of the scientists in the historic cases. Not science, not scientists, but the network of history, values, beliefs, and structures that comprises the American political system provided protection.

Human Life Bill

A flurry of interest occurred over another effort to legislate scientific truth in 1981. The ideological program of Reagan and other conservatives who had just gained office included a promise to pass antiabortion legislation. The new administration watched approvingly as a bill was introduced by Senator Jesse Helms of North Carolina that would recognize the right of the unborn to protection under the Fourteenth Amendment to the Constitution. The Fourteenth Amendment prohibits states from denying persons life without due process of law. The Helms bill explicitly included the unborn under this protection and designated abortion as a denial of due process.

The abortion controversy has long divided Americans, and there is no indication that the matter will soon be resolved. Apart from the merits of the arguments of pro- and antiabortionists, however, the Helms bill would have imposed a scientific truth on American society. The title of the proposed law was:

"A Bill to provide that human life shall be deemed to exist from conception." Section 1 stated: "The Congress finds that present day scientific evidence indicates a significant likelihood that actual human life exists from conception." The bill's text treated this "likelihood" as a certainty.

By declaring the beginning of human life to be at conception, the intention of the bill's supporters was to prohibit abortions for any reason at any stage of pregnancy, and defying the law could subject an offender to charges of murder. By establishing its definition of the beginning of life, the bill would amount to a political imposition of scientific truth, as in the Galileo, Lysenko, and Nazi experiences. This assault on science, if successful, would deserve the historic-case analogies that have so frequently and inappropriately been applied to other scientific activities in the United States.

The political machinery that has protected the citizenry from imposed scientific truth in the past quickly began to respond. Within months of the bill's introduction, the effort to mandate scientific fact seemed doomed. Although the abortion issue continues to simmer, institutional obstacles to legislate scientific truth have caused antiabortionists to turn to other methods. A review of the incident is instructive.

The impetus for enacting legislation to define when human life begins came from a 1973 Supreme Court decision, *Roe* v. *Wade.* The decision enraged antiabortionists. In removing restrictions that states had placed on abortions, the Court included a phrase that touched on the difficulty of legislating or adjudicating questions of scientific truth:

> We need not resolve the difficult question of when life begins. When those trained in the respective disciplines of medicine, philosophy and theology are unable to arrive at any consensus, the judiciary, at this point in the development of man's knowledge, is not in a position to speculate as to the answer.[33]

The difficulty in establishing such a definition was only one of several justifications raised by the Court for arriving at its liberalizing opinion. Antiabortionists, however, countered that the decision "flies in the face of well-established scientific fact."[34] After the 1980 elections, Republicans gained control of the Senate and right-wing leaders lost little time in pressing their cause. Senator Helms introduced his antiabortion bill on the first day of the new congressional session, January 19, 1981. Three months later the Subcommittee on the Separation of Powers began hearings on the bill. Its chairman was John East of North Carolina, a protégé of Helms who had just been elected to the Senate.

East tried to provide a forum that was strikingly biased toward the bill. During the initial sessions, eight scientists and physicians testified, seven of whom favored the bill and its human life definition. They offered straightforward support, typically expressed by the first witness, Dr. Jerome Lejeune, a professor of genetics in Paris. He declared that human beings start their existence at the moment of conception.[35] The view was echoed by every succeeding witness except the last, Dr. Leon Rosenberg, a geneticist at Yale, who appeared on the final day of the hearings.

Dr. Rosenberg argued that the question of when human life begins is not a scientific one but rather is philosophic and religious. He pointed out that reputable observers have argued alternatively that human life begins when brain function appears, when the heart begins to beat, when a recognizable human form appears in miniature, when the fetus can survive outside the uterus.[36] He made clear that he was not addressing the merits of the abortion argument but only the enormous error in trying to legislate scientific truth. He offered a gentle but powerful indictment of the seven witnesses who appeared previously:

> If I am correct in asserting that the question of when actual life begins is not a scientific matter, then, you may ask, why have so many scientists come here to say that it is? My answer is that scientists, like all people, have deeply held religious feelings to which they are surely entitled. In their remarks at these hearings, however, I believe that those who have preceded me have failed to distinguish between their moral or religious positions and their professional scientific judgments.[37]

Finally, Dr. Rosenberg questioned the fairness of the hearings:

> In these hearings I think I am correct that you have heard perhaps seven scientists and medical people say that they can accept Section 1 [of the Helms bill] and I am the only one who has appeared before you to say that I do not believe that they can.

> I hope that you and the rest of the subcommittee would not take the ratio of 7 to 1 to reflect the scientific opinion of the American scholar. If you want to know more about how American scientists feel on this issue, I would hope you will continue to seek additional opinion. . . .
>
> I would think a great majority of the American clinicians and scientists in this country would support my side of this argument despite my distinct minority before you in the past 2 days.[38]

This was not the first criticism directed at Senator East's organization of the hearings. The two Democratic members of the five-man subcommittee had refused to attend the sessions. One of them, Senator Max Baucus of Montana, had complained that the Democratic minority counsel had been barred from questioning the witnesses, and that Senator East had refused to schedule witnesses selected by the Democrats.[39] Even some conservatives expressed disquiet about Senator East's approach and the bill's attempt to define scientific truth. Senator Orrin Hatch of Utah, a Republican member of the subcommittee, announced that "I am second to none in my opposition to abortion, but I am equally committed to sound constitutional principles." He questioned whether Congress had the authority to define when human life begins.[40]

Before the hearings were over, Senator East had yielded to the crescendo of criticism and announced that sessions would be scheduled in the future to hear "all points of view."[41] This was a dramatic turnabout from what had been intended as an unequivocal display of support for the bill. Momentum against the human-life definition continued to grow. The National Academy of Sciences passed a resolution asserting the impossibility of saying in clear-cut scientific terms when human life begins.[42] Other organizations issued similar statements.[43]

Senator East had been pressed to the defensive. He scheduled additional hearings the following month, and the subcommittee listened to a more equitable representation of opinion. Most witnesses affirmed that the moment when human life begins could not be scientifically determined. It had become clear that the bill would go nowhere.

In an effort to salvage their effort, the proabortionist majority of three on the subcommittee reworded the bill, and dropped all reference to science or scientific evidence. As a result, the bill's first section appeared even more preposterous: "The Congress finds that the life of each human being begins at conception." Now, without even the pretense of appeal to scientific evidence, the bill was saying that Congress itself was the final scientific authority.[44]

Although the revised bill was voted out of Senator East's subcommittee by three to two, most observers felt it was unconstitutional. The full Judiciary Committee failed to consider it, effectively dooming its chances. Meanwhile foes of abortion began to weigh alternative routes. By early 1982 several proposals were before Congress. The one seen as drawing the largest range of support was a proposed constitutional amendment by Senator Hatch that would repeal the 1973 Supreme Court decision. The amendment was simply worded: "A right to abortion is not secured by this Constitution. The Congress and the several states shall have concurrent power to restrict and prohibit abortion: *Provided,* That a provision of a law of a State which is more restrictive than a conflicting provision of a law of Congress shall govern." In March 1982 the Judiciary Committee endorsed the Hatch amendment by a vote of ten to seven, thus setting the stage for full Senate consideration.[45]

While the abortion issue remains as divisive as ever, the issue of imposed scientific truth has withered. During consideration of the Hatch amendment, members of the Judiciary Committee made clear their uneasiness with the Helms bill's effort to define life. In its report supporting the Hatch amendment, the committee explicitly would "not attempt to define human life," and "takes no position on the question of when human life begins."[46]

In mid-1982, Senator Helms and a few other antiabortionists continued to press for a politically imposed definition of the beginning of life. Observers commonly saw this as a futile gesture.[47] Whether for political expediency or out of respect for scientific truth, foes of abortion, like the National Right to Life Committee and the U.S. Catholic Conference, turned from the human life bill and endorsed the Hatch amendment. The mix of structural checks and balances, open debate, and expression of competing interests once again stood as protector of the citizenry from politically imposed scientific truth.

DEFINING DEATH

An extension of the question about defining when life begins involves a statutory effort to define death. Because of enormous improvements in life-saving technologies, the effort has taken on increased medical and legal significance. Care of the terminally ill, decisions about when to remove vital organs for transplantation, and questions about when to suspend life-support systems are issues that remain unsettled.

As in trying to establish when life begins, defining "death" is an arbitrary exercise. It has variously been designated as the cessation of respiration, of heartbeat, or of brain function. A national commission studied the issue and proposed that states adopt uniform statutory laws. The commission's proposed definition has gathered general support: "An individual who has sustained either (1) irreversible cessation of circulatory and respiratory functions, or (2) irreversible cessation of all functions of the entire brain, is dead."[48]

By establishing a law that defines death, scientific truth is not being imposed. Rather, a situation becomes codified whereby technology and human values are joined in a determination that nothing more can or should be done for a moribund person. Mary Segers suggests that it is inappropriate even to use the term "define" about the condition. She would say that "science *identifies* [her italics] the characteristics of morbidity, the point beyond which investment of currently available medical resources . . . will not result in physical recovery or even the independent physiological functioning of the patient."[49]

In defining death one can no more claim absolute scientific authority than in pinpointing when human life begins. In both cases, an inescapable arbitrariness joins current scientific wisdom with human values. The absurdity of "scientifically" defining either life or death becomes clear when the standards of the Helms bill are brought to a definition of death. If human life begins the moment the first cell (zygote) is created, as Helms would have it, then death should not be said to occur until the last cell of a moribund person perishes. Apart from the technical implausibility of examining every cell in a moribund individual, the proposition is untenable on its face.

It serves further to demonstrate the futility of parading such definitions as scientific truth. Precise definitions of life and death are fabrications by humans, based on human values, for human purposes.

No one can "prove" that political impositions like those in the historic cases can never occur in the United States. Such impositions would require concurrence of the President, Congress, the Supreme Court, and an acquiescent public. But as long as American political structures are federated, and openness remains a central value of the system, mandated scientific truth seems a virtual impossibility. The lessons of fundamentalism, creationism, and the human life bill support the proposition. Even when endorsed by political leaders, public opinion, or outspoken interest groups, efforts to impose these issues as scientific truth made little headway.

Notes

1. Kuhn's book was first published in 1962. A second enlarged edition appeared in 1970. Thomas S. Kuhn, *The Structure of Scientific Revolutions,* 2d ed., enlarged (Chicago: University of Chicago Press, 1970). Representative views of historians and philosophers of science about issues raised by Kuhn are in Frederick Suppe, ed., *The Structure of Scientific Theories,* 2d ed. (Urbana, Ill.: University of Illinois Press, 1977).
2. J. Robert Oppenheimer, *The Open Mind* (New York: Simon and Schuster, 1963), pp. 92–93.
3. The Scopes trial and related issues are recounted in L. Sprague de Camp, *The Great Monkey Trial* (Garden City, N.Y.: Doubleday and Co., 1968); Jerry R. Tompkins, ed., *D-Days at Dayton* (Baton Rouge, La.: Louisiana State University Press, 1965); John T. Scopes and James Presley, *Center of the Storm, Memoirs of John T. Scopes* (New York: Holt, Rinehart and Winston, 1967). Documents pertaining to the case are in Sheldon Norman Grebstein, ed., *Monkey Trial, The State of Tennessee vs. John Thomas Scopes* (Boston: Houghton Mifflin Co., 1960).
4. Roger Baldwin, "Dayton's First Issue," in Tompkins, p. 62.
5. Gail Kennedy, ed., *Evolution and Religion* (Boston: D. C. Heath and Co., 1957), p. ix.
6. Grebstein, pp. 81–82.
7. *The New York Times,* January 17, 1927, p. 1.
8. Fay-Cooper Cole, "A Witness at the Scopes Trial," *Scientific American,* vol. 200, no. 1 (January 1959), p. 130.
9. Watson Davis, "The Men of Science," in Tompkins, pp. 72–73.
10. Judith V. Grabiner and Peter D. Miller, "Effects of the Scopes Trial," *Science,* vol. 185, no. 4154 (September 6, 1974), pp. 832–37.

11. Some teachers taught evolution to their classes on the lawn outside of school, so they could not be accused of teaching it *in* the public school. Others would announce they could not discuss the chapter on evolution in the textbook, and were convinced their students then read the chapter more assiduously than if it had been assigned. See de Camp, *The Great Monkey Trial,* p. 487; and Davis, p. 73.

12. An antievolution law in Arkansas was challenged in 1962 by a young biology teacher named Susan Epperson. In 1968, the U.S. Supreme Court found the law unconstitutional, confirming legally what had been the factual understanding for forty years. See L. Sprague de Camp, "The End of the Monkey War," *Scientific American,* vol. 220, no. 2 (February 1969), pp. 15–21.

13. More than one hundred reporters covered the trial, and a record number of dispatches were sent around the world. *The New York Times,* July 22, 1925, p. 2.

14. Baldwin, p. 61.

15. Bert James Loewenberg, "The Reaction of American Scientists to Darwinism," *The American Historical Review,* vol. 38, no. 4 (July 1933); W. C. Curtis, "The Evolution Controversy," in Tompkins, p. 75.

16. Kirtley F. Mather, "Geology and Genesis," in Tompkins, pp. 88–91.

17. de Camp, "The End of the Monkey War," p. 19. Twelve scientists went to Dayton to work with the defense. Sixteen others had signified their willingness to go if called upon. See *Science,* vol. 62, no. 1597 (August 7, 1925), p. 130.

18. Curtis, p. 76. The American Civil Liberties Union sponsored the defense. Two years after the Scopes trial a proposed antievolution bill in Minnesota was greeted with indifference by the scientific community. "No scientist as such appears against the bill or for it," wrote a scientist named Frederick Sandeson in justification of the scientists' attitude, because the issue "is political when not theologic." *Science,* vol. 65, no. 1688 (May 6, 1927), p. 447.

19. Grabiner and Miller, p. 836.

20. *Ibid.*

21. These statements appear in numerous publications by the Institute for Creation Research, including its series of pamphlets titled *Acts and Facts,* and *Impact.* The Institute, located in El Cajon, California, is the research division of Christian Heritage College. It seeks to project itself as the prime scholarly and scientific arm of the creationist movement. See Dorothy Nelkin, *Science Textbook Controversies and the Politics of Equal Time* (Cambridge, Mass.: The M.I.T. Press, 1977), pp. 68–70.

22. Henry M. Morris and Gary E. Parker, *What Is Creation Science?* (San Diego, Cal.: Creation-Life Publishers, 1982), p. 251.

23. *Ibid.,* p. 239.

24. *Ibid.,* pp. 116–29.

25. The Arkansas case is comprehensively examined in a special section of *Science, Technology, and Human Values,* vol. 7, no. 40 (Summer 1982). In November 1982, a federal judge ruled that the Louisiana law violated the state's constitution because the legislature had no authority to dictate how a subject must be taught in the public schools. Constitutional experts were convinced that if the decision were appealed, the law would ultimately be

held unconstitutional by the Supreme Court. *The New York Times,* November 23, 1982, p. A-23.

26. Comment by Maxine Singer in *Science,* vol. 214, no. 4521 (November 6, 1981), p. 635.

27. Comment by William Mayer in *ibid.*

28. Judge Overton's opinion is reproduced in *Science, Technology, and Human Values,* vol. 7, no. 40 (Summer 1982), pp. 28–42; and Jeffrie G. Murphy, *Evolution, Morality, and the Meaning of Life* (Totowa, N.J.: Rowman and Littlefield, 1982), pp. 117–48.

29. Joel Gurin, "The Creationist Revival," *The Sciences,* vol. 23, no. 4 (April 1981), p. 16.

30. An NBC News poll taken in late 1981 found that 76 percent of the respondents felt that public schools should teach both evolution and creationism. *The New York Times,* November 18, 1981.

31. From Judge Overton's opinion. Also see *Science,* vol. 215, no. 4531 (January 22, 1982), p. 384.

32. Telephone interview, July 1982, anonymity requested.

33. U.S. Congress. Senate. Report of the Committee on the Judiciary on S. J. Res. 110. *Human Life Federalism Amendment* (Washington, D.C.: Government Printing Office, June 8, 1982), p. 32.

34. James L. Buckley, "A Human Life Amendment," *The Human Life Review,* vol. 1, no. 1 (Winter 1975), p. 8.

35. U.S. Congress. Senate. Hearings before the Subcommittee on Separation of Powers of the Committee on the Judiciary on S. 158. *The Human Life Bill.* April 23, 24; May 20, 21; June 1, 10, 12, 18, 1981 (Washington, D.C.: Government Printing Office, 1982), pp. 7–10.

36. *Ibid.,* pp. 49–50.

37. *Ibid.,* p. 51.

38. *Ibid.,* p. 60.

39. *The New York Times,* April 24, 1981, p. A-16.

40. Senate Hearings on *The Human Life Bill,* p. 4.

41. *The New York Times,* April 25, 1981, p. 7.

42. *Ibid.,* April 29, 1981, p. A-25.

43. Senate Hearings on *The Human Life Bill,* p. 1089.

44. Both versions of the bill are reproduced in Hearings before the Senate Subcommittee . . . on S. 158, pp. 1117–24.

45. Congressional approval of a Constitutional amendment requires a two-thirds majority in each House, as opposed to a simple majority for ordinary legislation. Even so, by avoiding the "scientific truth" issue, the amendment was given a better chance to pass.

46. Report on the *Human Life Federalism Amendment,* pp. 33, 64.

47. *The New York Times,* March 10, 1982, p. A-26. During a Senate debate on the various antiabortion bills in August 1982, Senator Helms recognized the futility of his approach and withdrew references to the beginning of life from his bill.

48. U.S. Government. President's Commission for the Study of Ethical Problems in Medicine and Biomedical and Behavioral Research, *Defining Death* (Washington, D.C.: Government Printing Office, July 1981), p. 73.

49. Mary C. Segers, "Can Congress Settle the Abortion Issue?" *The Hastings Center Report,* vol. 12, no. 3 (June 1982), p. 22.

7

Unethical Experimentation

We have thus far sought to sustain two broad propositions. First, neither science nor scientists are principal protectors against political impositions of scientific truth. The nature of a society's political institutions stands as the fundamental determinant. Second, despite perceptions by many to the contrary, Americans need not fear that regulating scientific activity would lead to perversions of science as occurred in seventeenth-century Italy, Stalin's Russia, or Nazi Germany. In consequence, critics who associate governmental regulation of scientific inquiry in the United States with the historic cases of scientific perversion obfuscate rather than clarify.

In denigrating the need for political restraint of some scientific activities, critics ignore the potential for American political institutions to act as protectors. Restraint of science, they say, is always bad, never appropriate. In fact, restraint of scientific activity as a result of imposed truth by political authority is not legitimate, but restraint of scientific activity that endangers the public should always be deemed appropriate. These two features are often blurred by careless analogy.

Failure to differentiate between the two has sometimes inhibited the government from acting to protect the population from dangers born of scientific activity, and this reluctance to institute protective machinery has contributed to unnecessary risks to public safety. This chapter will examine how the lack of institutional oversight was responsible for harming people who were subjected to unethical experimentation. At the same time, efforts to remedy past errors demonstrate the potential within the political system to protect the public. The citizenry is better served because the government, though belatedly and tentatively, has restrained this avenue of scientific inquiry.

During the 1970s the public learned that governmental agencies had been responsible for experimentation on unwitting subjects for decades. The experiments had endangered the subjects' lives. Three celebrated cases reveal the nature of this research and its relationship to the political system. In all three cases governmental regulation initially had been egregiously lacking. Public disclosure set in motion institutional responses, and protection of the population was enhanced. The three cases prompted political restraint of scientific inquiry that proved necessary and salutary. After reviewing the cases, corrective efforts that they generated will be discussed, as well as why additional steps are still warranted.

The Tuskegee Syphilis Study

Beginning in 1932, the United States Public Health Service administered placebos to several hundred black men in Tuskegee, Alabama under the guise of treating them for syphilis.[1] During the next forty years the naive subjects returned periodically for purported treatment. In fact, they were being observed in an effort to determine the effect of untreated syphilis on physical and mental health.

By 1952, 40 percent of those who went untreated had died, compared to 20 percent who had received treatment.[2] Nevertheless, the study continued. Not until 1966 did a public health adviser question the ethics of the experiment. His inquiries were shunted aside, however, and he later spoke to a journalist about the study.[3] Following press reports the study was ended in 1972.

Soon after, an investigative panel was appointed by the Department of Health, Education, and Welfare, and it repudiated the study the following year.[4] In its final report, the panel concluded that the research had been "ethically unjustified" and "scientifically unsound."[5] As a result of a law suit, the federal government in 1974 agreed to pay $9 million to the survivors or their heirs. By 1979 survivors or heirs of all but seventeen had been located and compensated.[6]

CIA Mind-Control Experiments

Three years after the Tuskegee study was exposed, the country learned that the Central Intelligence Agency had also been experimenting on unwitting subjects. A commission appointed by President Ford in 1975 to investigate CIA activities in the United States disclosed that during the fifties the agency was responsible for administering LSD to unsuspecting subjects. The extent of CIA drug programs was not recognized until two years later, when a team of journalists sifted through two thousand agency documents obtained under the Freedom of Information Act. Mind-control experiments with a variety of drugs had been performed under CIA auspices for more than twenty-five years at a cost of $25 million.[7]

There was little apparent concern about the health of the subjects beyond the immediate purposes of the experiments. Two people died during the drug tests and unknown hundreds may have suffered long-term problems as a consequence of the mind-altering tests. The individuals who conducted the tests often were not qualified scientists, nor were subjects observed for more than a few hours after the tests.[8]

One of the deaths occurred in 1953 when Frank Olson was administered a glass of liqueur containing LSD. He developed a psychotic reaction and a week later committed suicide. The experiments were continued because, according to a scientist associated with the testing program, the physicians advising the agency were unable to find an "absolute" connection between the LSD and the suicide.[9] In recognition of its responsibility, however, in 1975 the government awarded $750,000 in compensation to Olson's survivors.

In 1977, following initial newspaper disclosures, CIA director Stansfield Turner announced that additional records of tests had been found and were being made public. While testifying before a Senate committee, Turner acknowledged that the testing took place on "criminal sexual psychopaths confined at a state hospital" and on terminal cancer patients, all of whom were unaware of the dangers involved. He deplored these programs and said that all CIA drug testing had been ended.[10]

The Army Germ Warfare Tests

One morning in December 1976 Edward Nevin III stood on a commuter platform awaiting the train to his San Francisco law office. As he browsed through a newspaper he came across an article that stunned him. The incident marked the beginning of an odyssey that would lead to a suit against the government for $11 million.[11]

The article disclosed that the Army had conducted experiments in 1950 to test San Francisco's vulnerability to germ warfare attack. A reporter had learned about the project from a scientist who had information about the secret experiments twenty-six years earlier. The article revealed that about a month after the test "one man was dead and five other patients were infected at a local hospital by the same kind of bacterium used in the test."[12] The man who had died had been admitted to Stanford University Hospital for prostate surgery. He was otherwise in good health. His name was Edward Nevin. Now the younger Mr. Nevin learned for the first time about the peculiar circumstances that surrounded his grandfather's death.

The newspaper article began a series of revelations about the most awesome experimentation conducted by an agency of American government on unwitting subjects. At congressional hearings in 1977, the Army acknowledged that it had secretly tested the dispersion of bacteria and chemicals during a twenty-year period beginning in 1949. A report made public by the Army cited 239 simulated germ warfare tests that had been conducted in populated areas throughout the United States.[13]

One of the microorganisms most frequently used was *Serratia marcescens.* Reports had appeared before 1950 that questioned the safety of the bacteria, and by the end of the 1950s widespread evidence had accumulated implicating *Serratia marcescens* as a cause of infection and death. Nevertheless, the Army continued to use it over large populations during the next decade. The Assistant Secretary of the Army for Research and Development, Edward A. Miller, testified at the congressional hearings that the tests were not ended until 1970.[14]

The consequences of the earliest tests should have raised doubts about safety from the outset. In 1950 a Navy ship secretly sprayed a smog of serratia over the San Francisco Bay area during a period of two days. The Army determined that nearly every one of the 800,000 people in San Francisco unknowingly inhaled five thousand or more fluorescent particles mixed with the serratia.[15] Four days later a patient at Stanford University Hospital was found to be suffering from infection caused by *Serratia marcescens,* the first such case ever recorded at the hospital. During the next five months, ten more patients at the hospital were found to have been infected by serratia, one of whom, Edward Nevin, died.

Since serratia infections had rarely been reported elsewhere, and never before at Stanford Hospital, doctors at the hospital were baffled. They knew nothing about the Army tests. When the Pentagon learned of the outbreak of serratia infections it did not inform anyone at Stanford about the tests, but convened a secret committee of four scientists to assess the situation. The scientists concluded that the appearance of the cases at about the time of the bacterial warfare tests "appeared coincidental." They recommended that the use of *Serratia marcescens* "as a simulant should be continued, even over populated areas, when such studies are necessary to the advancement of the Biological Warfare program."[16]

During the next two decades, huge amounts of serratia and other bacterial and chemical agents were secretly sprayed over populated areas, including Washington, D.C., Key West, and Panama City. In 1966 bacteria were pumped into the New York subway system.[17]

In the late seventies San Francisco physicians began to report *Serratia marcescens* infections of the heart among addicts who injected themselves with heroin and other drugs. Since no other part of the country had experienced serratia infections as frequently as San Francisco, including areas with as many addicts, physicians suspected that the Army tests in 1950 may have seeded the area.[18] An Army spokesman in 1977 denied any possible relationship.[19]

Lawrence K. Altman, a physician who surveyed the issue in an article for *The New York Times,* wrote: "So little is known about this phenomenon, and the biological aspects of the situation are so complex, that it appears virtually impossible to make a scientific judgment about whether the biological warfare tests caused the hospital cases."[20]

Meanwhile, Edward Nevin's children, grandchildren, and great-grandchildren, in their suit against the Army, charged that Nevin's death resulted from exposure to the bacteria released during the tests. The non-jury trial was conducted in March 1981 before Federal District Court Judge Samuel Conti. The thirty-one-year gap between Mr. Nevin's death and the trial created many uncertainties—lost evidence, destroyed hospital records. The judge ruled against the plaintiffs, and the case is now on appeal.

Scientific and Political Responses

The three cases of experiments on unwitting subjects were reprehensible by any standard of ethics. In each case people's lives were endangered. Decisions by scientists appeared to disregard the health and safety of the experimental subjects. In 1969, when a group of physicians evaluated the Tuskegee study for the United States Public Health Service, they advised that it be continued. When the question was raised about the propriety of withholding treatment from the unsuspecting syphilitic patients, one consultant said: "You will never have another study like this; take advantage of it."[21]

Another Public Health Service physician acknowledged that "nothing learned will prevent, find, or cure a single case of infectious syphilis or bring us closer to our basic mission of controlling venereal disease in the United States." Yet he too urged that the study be continued "along its present lines."[22] This view prevailed until 1972 when widespread media coverage forced the cessation of the study.

In the case of the CIA mind-control experiments, the death of a victim who had unwittingly taken LSD appeared suspicious. As mentioned, because the connection between LSD and the victim's death was not "absolute" according to scientists who reviewed the incident at the time, the experiments were continued.

Similarly, four experts were requested by the Army to examine the possible relationship between the blanketing of the San Francisco area with *Serratia marcescens* and the subsequent cases of infections at Stanford University Hospital. Despite the fact that serratia infections had never before been reported at the hospital (and rarely anyplace else), and that one person died from heart and blood complications brought on by serratia, the Army concluded that "the association of SM infections with the San Francisco tests appeared coincidental."[23] The four consultants, described as "eminent scientists" by the Army, urged that the use of serratia in tests over populated areas be continued.[24]

These surely were ethical contraventions, and the fact that they were conducted for decades reveals a terrible weakness in the political system that sustained them. Paradoxically, they help point out strengths of the system as well.

First, the experiments were not part of a national scheme that condoned research on unwitting subjects, nor a consequence of imposed scientific truth, as in the Nazi case. Rather they were secretly conducted within isolated units of the government and were not statements of national policy.

Second, while the experiments were conducted under government auspices, condemnation was prompted by efforts from within the government. In the Tuskegee case, a public health adviser familiar with the study was the catalyst. In the instance of the CIA, it was an investigative committee appointed by the President. For the Army experiments, a newspaper reporter obtained information from a scientist who had been a consultant to the government. When information about each case reached sources outside of the agencies involved, Congress and the press investigated further. Access to agency documents was enhanced by passage of the Freedom of Information Act in 1974. Through a variety of mechanisms within the political system, public exposure to the issues was broadened.

Third, the government acknowledged its errors through official spokesmen and financial compensation was paid to victims or survivors. (This has not been true in the germ warfare case, which, as mentioned, is still before the courts.) Remedial actions prompted by the exposure of these incidents will help minimize the chances that similar experimentation on unwitting subjects will take place in the future. Codes of ethics dealing with experimentation involving humans were issued by several professional associations.[25] Federal regulations for safeguards in human experimentation were devised, including the creation of a National Commission for the Protection of Human Subjects and mandatory institutional review boards at colleges, hospitals, and other research institutions.[26]

A new discipline, bioethics, emerged during the 1970s. Institutes were formed to study a range of bioethical issues and have become clearinghouses of information. The Institute of Society, Ethics and the Life Sciences at the Hastings Center (founded in 1969) and the Kennedy Institute of Ethics at Georgetown University (established in 1972) are two of the most prominent. They are frequently consulted by private groups and government officials on ethical questions about experimentation and other life science issues. In little more than a decade, policy-makers and research institutions have become sensitive to bioethical questions that few had thought about earlier.

Thus, the revelation that scientists in several governmental agencies had conducted unethical experiments helped spawn a series of institutional responses to protect the citizenry from continuation of such practices. The federal effort to oversee experimentation on human subjects derives principally from the National Research Act of 1974. While the National Commission was established in addition to other protective institutions, protection remains incomplete and activities since passage of the act have demonstrated both its benefits and deficiencies.

The National Commission was succeeded in 1978 by the President's Commission for the Study of Ethical Problems in Medicine and Biomedical and Behavioral Research. Despite the expansive title, the President's Commission has little power. It has no regulatory authority; it can only study and make recommendations. In December 1981, the President's Commission published an extensive report entitled *Protecting Human Subjects.* The report summarized the existing status of protection.

On the positive side, all federal agencies involved with research on humans now have institutional review boards (IRBs). There exists a sensitivity about the rights of human subjects that did not exist before the 1970s, but, as the report discloses, regulations regarding protection vary from agency to agency. The responsibilities and effectiveness of IRBs are not uniform. Even the definition of scope and function of the

IRBs remains ambiguous.[27] For example, the regulations and the IRB system created by the Department of Health and Human Services (cited as an excellent model by the President's Commission) provide for sanctions in cases of noncompliance with the Department's regulations. Penalties include suspension of individuals from further financial assistance within the department. But in the Department of Agriculture, sanctions for noncompliance are not specified.[28] Agencies are inconsistent in their efforts to ensure that their regulations are being adhered to. Some agencies do not "really do much of anything to know how well their regulations are working," according to Alexander M. Capron, executive director of the President's Commission.[29] One of the tasks the President's Commission had set for itself was to draw a uniform set of regulations and urge all federal agencies to adopt them.

Two other problems exist that are not receiving active attention. First, regulations apply only to research involving federal funds. Private institutions may operate research projects on humans without review boards or formalized protective machinery. Second, some forms of research that may endanger the public are apparently not covered by existing regulations. Mr. Capron is confident that the public is now protected from experiments like the Tuskegee syphilis study and the CIA mind-control experiments. But he thinks that the Army germ warfare experiments could be repeated "under present rules and general perceptions of proper standards." Although he would be unhappy if these tests were resumed, he thinks the possibility exists because the Army's definition of research might not include spraying millions of people with "harmless" bacteria. They consider research as something "more directly controlled and medical," Mr. Capron says. The exposed population would not have to give informed consent before being sprayed, according to this interpretation.

Besides the loopholes in existing protective procedures, the President's Commission goes out of existence at the end of 1982, according to the bill that established it four years earlier.[30] Because of the ever-present danger of uncontrolled experimentation, the commission should be made permanent and given regulatory authority. Subjecting unwitting citizens to experimentation with bacteria should be prohibited. The Army and every other agency should be accountable for experimentation on human subjects to an authority beyond themselves.

In sum, a review of unethical experimentation sponsored by governmental agencies demonstrates both weaknesses and strengths in the government's efforts to protect its citizens. Reluctance to restrain scientific activity, because of what some consider a scientist's First Amendment rights, contributed to excesses of the past. Although protection against unethical experimentation is not yet optimal, opportunities to correct deficiences remain available within the system.

The characteristics of the American political system that differentiate it from the systems in the historic cases provide American society with its best protection against both imposed truth and unethical experimentation. A political tradition of accommodation rather than violence to resolve conflicts, a political culture that invites general participation and openness, and political structures based on federalism, checks, and balances rather than centralized direction of power, are all crucial features of the American experience.

These virtues have helped resolve the challenges wrought by the attempted imposition of scientific truth in the form of antievolution and the human life bill. They have been essential as well to the improvement in protecting citizens against unethical experimentation.

Notes

1. Unlike the Nazi experiments, the American victims were not designated by a national doctrine based on scientific "truth." Nevertheless, racism obviously was central to the selection of blacks as subjects. Allan M. Brandt, "Racism and Research: The Case of the Tuskegee Syphilis Study," *The Hastings Center Report*, vol. 8, no. 6 (December 1978), pp. 21–29. Also, James H. Jones, *Bad Blood* (New York: The Free Press, 1981).
2. Richard M. Restak, *Premeditated Man, Bioethics and the Control of Future Human Life* (New York: Penguin Books, 1977), p. 112.
3. *Ibid.*, p. 187.
4. U.S. Department of Health, Education, and Welfare, Public Health Service, *Final Report of the Tuskegee Syphilis Study Ad Hoc Advisory Panel,* April 28, 1973, pp. 6–8.
5. *Ibid.*, pp. 7–8.
6. *The New York Times,* July 29, 1979, p. 26.
7. *Ibid.*, August 2, 1977, pp. 1, 16.

8. U.S. Congress. Senate. Joint Hearing before the Select Committee on Intelligence and the Subcommittee on Health and Scientific Research of the Committee on Human Resources. *Project MKULTRA, the CIA's Program of Research in Behavioral Modification,* August 3, 1977 (Washington, D.C.: Government Printing Office, 1977), pp. 66, 80.

9. Testimony by Dr. Sidney Gottlieb before U.S. Congress. Senate. Hearings before the Subcommittee on Health and Scientific Research of the Committee on Human Resources, *Human Drug Testing by the CIA, 1977,* September 20 and 21, 1977 (Washington, D.C.: Government Printing Office, 1977), pp. 185–86.

10. Joint Hearing. . . . , *Project MKULTRA. . . .* , pp. 7–8, 16.

11. I am grateful to Edward Nevin for sharing with me his recollections and documents that he obtained from the Army pursuant to his suit against the government.

12. Articles originally appeared in *Newsday* beginning November 21, 1976, and were carried in newspapers across the country. See U.S. Congress. Senate. Hearings before the Subcommittee on Health and Human Resources of the Committee on Human Resources, *Biological Testing Involving Human Subjects by the Department of Defense, 1977,* March 8 and May 23, 1977 (Washington, D.C.: Government Printing Office, 1977), pp. 141–42.

13. *Ibid.*, pp. 125–40.

14. *Ibid.*, p. 16.

15. *Special Report No. 142,* U.S. Army, Physical Defense Division, Biological Department, Chemical Corps, January 22, 1951.

16. *Information for Members of Congress: U.S. Army Activities in the U.S. Biological Warfare (BW) Program,* Office of the Secretary of the Army. March 8, 1977, p. 5.

17. Hearings. . . . , Biological Testing. . . . , pp. 125–40; *The New York Times,* March 9, 1977, p. A-1; March 13, 1977, p. 26; March 21, 1977, p. 21.

18. John Mills, M.D. and Denis Drew, M.D., "*Serratia marcescens* Endocarditis: A Regional Illness Associated with Intravenous Drug Abuse," *Annals of Internal Medicine,* vol. 84, no. 1 (January 1976).

19. *The New York Times,* March 13, 1977, p. 26.

20. *Ibid.*

21. Dr. J. Lawton Smith, quoted in Brandt, p. 26.

22. Dr. James B. Lucas, quoted in *ibid.*

23. *Information for Members of Congress. . . .* , p. 5.

24. *Ibid.*

25. Some associations, including the American Anthropological Association, American Psychological Association, and American Sociological Association, adopted codes in the early seventies prior to full disclosure of the incidents described here. After public disclosure, many more organizations adopted codes. The 1979–80 edition of the *Bibliography of Society, Ethics and the Life Sciences,* (New York: Institute of Society, Ethics and the Life Sciences, 1978), p. 7, lists thirty-one references under "Codes of Professional Ethics," twenty-three of which were established in the seventies.

26. The National Commission for the Protection of Human Subjects of Biomedical and Behavioral Research, review boards, and the legislated com-

mitment to protect human subjects arose from the National Research Act of 1974. Bibliographies about these and related subjects are in *ibid.*, pp. 30–38.

27. U.S. Government. President's Commission for the Study of Ethical Problems in Medicine and Biomedical and Behavioral Research, *Protecting Human Subjects* (Washington, D.C.: Government Printing Office, December 1981), pp. 35–36.

28. *Ibid.*, p. 126.

29. Quotations of Mr. Capron's in this and subsequent passages are from an interview conducted June 2, 1982.

30. Shortly before the end of 1982, the life of the commission was extended by three months to enable it to complete work in progress.

8

Nuclear Politics and a Comment About the Recombinant DNA Issue

The argument frequently posed by others that the restraint of any scientific activity in the United States risks leading to perversions on the order of those in 17th century Italy, Stalin's Russia, or Hitler's Germany has been found wanting. We have demonstrated that American political institutions are more suited to protect against imposed scientific truth than to perpetrate it, and that the system responded to information about unethical experimentation on humans by creating safeguards. In a different dimension of scientific activity, the ability of the political system to act appropriately has come under question.

Irreversible Hazards

Since World War II scientific and technological activities have induced irreversible conditions that threaten the welfare of much of the world's population, e.g., proliferation of nuclear waste, the spread of toxic and indestructible chemicals, the threat of recombinant DNA research. While the citizenry in a totalitarian or authoritarian political system is subject to political control against which it has no recourse, a citizenry in a democracy might now become subject to forces beyond its control as a consequence of certain scientific or technological activity. Even while its political institutions remain intact, catastrophic events could threaten the freedom and well-being of a large population. Millions could be endangered, for example, if radioactive waste contaminated underground water systems or if laboratory-created pathogens escaped to infest the general population. As political systems imposed truth and fact in the historic cases, so may the facts of irreversible science be imposed on masses of people. Totalitarian or authoritarian political systems induced the historic cases involving the perversion of science. Today, the greater threat to the world's population comes from scientific activity that may introduce irreversible hazards.

In examining this threat and how it might be confronted, the relationship of the political system to nuclear developments will be examined and then compared with the development of recombinant DNA policies. Each of these scientific enterprises has been characterized since its infancy by two preeminent features. Both have been perceived as capable of offering spectacular benefits to mankind, but also as fraught with irreversible dangers. While the response of the political system to nuclear development has been inadequate, as will be discussed, it has been creditable in the recombinant DNA controversy.

The nuclear story is a litany of how American political institutions were distorted to encourage the development of nuclear power. In the process, questions about safety were often disregarded. In the case of recombinant DNA research, traditional institutions that exposed controversial issues and encouraged debate became catalysts toward the creation of safeguards and popular confidence. The two issues are not the only examples of scientifically or technologically threatening activities, but they serve as well as any to demonstrate the strength and weakness of the American political system. They suggest how the system might best address future scientific activities that threaten the population with similar hazards.

The Failure of Nuclear Policy

For several days after the nuclear power reactor at Three Mile Island in Pennsylvania went amok in March 1979, members of the Nuclear Regulatory Commission engaged in a series of urgent meetings. Our national commitment to the expansion of nuclear power had been girded by assurances of safety by governmental agencies for thirty years. Now the five-member commission charged with regulating nuclear technology was at one with the rest of the nation behind a wall of ignorance.

The members were facing the possibility of a nuclear catastrophe that could kill thousands and make a portion of Pennsylvania uninhabitable indefinitely. The NRC chairman, Joseph Hendrie, despaired at the time that the commissioners were staggering blind men trying to make decisions.[1]

The incident highlighted the fact that the central weakness with our nuclear policy had not been scientific or technological but political. After promoting nuclear power for thirty years, we had only begun to debate in a national forum the risk of meltdowns, the escape of radioactive materials, and the disposal of nuclear waste. By any measure of reason, questions like these should have been answered before embarking on a program of nuclear expansion. Why didn't the debate occur earlier?

DISTORTING POLITICAL INSTITUTIONS

"The atom had us bewitched," recalled David Lilienthal, first chairman of the Atomic Energy Commission, in an interview. "It was so gigantic, so terrible, so beyond the power of imagination to embrace, that it seemed to be the ultimate fact. Our obsession with the atom led us to assign to it a separate and unique status in the world."[2]

In fact, there were three momentous periods in our nuclear history that called for an extended national debate. In each case, our political institutions failed to respond. They had been bent and tugged to accommodate our atomic obsession—that nuclear power *had* to be developed.

The first important moment occurred in 1946. In the aftermath of the atomic bomb explosions during the war, the government abandoned the usual structures for regulation and deliberation. An Atomic Energy Commission was established that, unlike any other regulatory agency, was assigned the contradictory responsibility of promoting the enterprise it was supposed to regulate.

At the same time, rather than subject nuclear affairs to the usual oversight arrangements involving separate committees in the Senate and the House, the rules were streamlined. In the dedication to a nuclear future, encumbrances that encourage debate, consideration, and reconsideration were set aside.[3] A single Joint Committee on Atomic Energy was empowered to oversee all matters pertaining to atomic energy. Every nuclear issue that came before Congress had to be heard by this committee, and only this committee. Rather than ask fundamental questions, the Joint Committee became a zealous advocate of nuclear power. Every member supported expanding nuclear energy, recalls Congressman Melvin Price, who served on it until its dissolution in 1977.[4] Amid the euphoria of the new atomic age no one seemed to doubt that nuclear development was inevitable. But the distortion of our political structure provided the underpinning that would lead to secrecy, deception, and the stifling of skepticism.

The second period of moment took place ten years later. Disclosures of memoranda obtained under the Freedom of Information Act reveal how single-minded and duplicitous the Atomic Energy Commission had become. Despite contrary evidence in its possession, the AEC was announcing publicly that radioactive fallout from nuclear tests in Nevada posed no danger to the people in the area. The mood of the AEC was reflected by Commissioner Thomas E. Murray's comment in 1955 in a memorandum intended for internal use: "We must not let anything interfere with this series of tests—nothing." Commissioner Willard Libby decided that people would have to "learn to live with . . . fallout," even if they did not know they were receiving it.[5]

Shocking revelations about the commission's methods continued to surface twenty-five years later. In 1980, a Defense Department technical adviser came upon recently declassified documents suggesting that the AEC had induced several scientists to deceive the public. Thousands of sheep in Utah and Nevada suddenly died in 1953 and the commission launched an inquiry. Scientists who investigated the deaths reported that they were caused by radioactive fallout following nearby nuclear weapons tests. The AEC concealed the reports and pressed the scientists by letters and visits from staff members to revise their conclusions. The scientists acquiesced and agreed not to challenge the official line that the sheep died of unknown causes.[6]

Two publicly reported incidents occurred in 1956 that should have signaled concern about the reliability of information coming from the nuclear establishment. Cancer specialists had charged that radioactive strontium-90, one of the lethal products in nuclear fallout, had contaminated large quantities of milk. Adlai Stevenson, then contending for the presidency against Dwight Eisenhower, related the charge and called for government action.

The White House rebuked Stevenson for his "amazing" and "incredible" statement. Lewis L. Strauss, the AEC chairman, declared: "The facts are directly counter to the charge. Mr. Stevenson's continuous efforts to frighten the public on the eve of election are not admirable."[7]

A congressional inquiry should have been in order, but the Joint Committee sat mute. The question of radioactive contamination became a non-issue, dismissed as an election campaign ploy.

During the same year word leaked that Strauss had tried to conceal a report that a proposed nuclear power plant to be built near Detroit could endanger the public. Even after the public disclosure of Strauss's effort, the AEC ignored the report and issued a construction permit. (Ten years later, as the plant was about to go into production, a harrowing accident recounted by John Fuller in *We Almost Lost Detroit* forced its permanent shutdown.[8])

Surely not every member of the AEC or others holding high office can be accused of venality. Like most of the citizenry, they had come to view a nuclear future as manifest destiny. They had become so psychologically committed to it that they apparently persuaded themselves that distorting facts could be justified in the interest of more quickly realizing that destiny. A few members of the Joint Committee expressed irritation with the AEC's cavalier approach. But instead of promoting a national debate about the wisdom of expanding nuclear power or even of existing regulatory arrangements, the Joint

Committee went ahead with the preparation of its legislative monument to nuclear power, the Price-Anderson Act of 1957.

Despite enormous government subsidies for nuclear development, private companies had been slow to invest. The Price-Anderson Act was intended to solve that by limiting the maximum liability in the event of an accident to $560 million, though reports acknowledged that damage could exceed $7 billion.[9] Consequently even if a company's negligence caused a major accident, it would be nicely shielded. A victim might receive compensation for less than 8 percent of his medical and burial expenses. Hardly anyone wondered about the logical contradiction now being raised with frequency: If nuclear power is as safe as proponents claim, why should it require special protection against liability for accidents?

In the mid-fifties the view that nuclear power was "our great sacred trust," as one Senator put it, went unchallenged.[10] Skeptical comments by an occasional outsider, even a presidential candidate, confronted an intolerant belief-system that was nourished by a burgeoning nuclear establishment. There simply was no level of government that would tolerate skepticism.

The third period of our nuclear history that should have prompted a national debate occurred in the late sixties. For the first time a few insiders who had been involved with our nuclear program began to raise questions.

David Lilienthal speaks of his experience in the mid-sixties when he urged more attention to safety in the nuclear power program. "I was treated as a traitor," he recalls. "Scientists and engineers joined in a campaign that Westinghouse put on to discredit my ideas."[11]

When two scientists who worked for the AEC, John Gofman and Arthur Tamplin, calculated in 1969 that radiation from nuclear reactions was more hazardous than had been assumed, they lost their jobs. They were condemned by spokesmen for the AEC and the nuclear industry. When their predicament was reported in the public media, rather than inquire into the implications of the scientists' criticisms, the Joint Committee, the only congressional committee empowered to oversee atomic matters, joined in condemning them.[12]

Alvin Weinberg, a physicist preeminently associated with the development of nuclear power, recalled a conversation he had around that time with "a very highly placed person in Congress. He said to me, 'Alvin, if you have concerns about nuclear safety, it's time for you to leave this business.' "[13] The theme appeared to be an inviolable principle of the nuclear establishment.

There were several reasons that skepticism by experts in the sixties, and the harsh response to it, went largely unnoticed by the public. In the first place the nation was riveted on the civil rights and Vietnam issues. People were reeling over the assassinations of John and Robert Kennedy and Martin Luther King. They were the transcendent national concerns. Energy sources seemed plentiful; at the end of the sixties nuclear power accounted for only 1 percent of the nation's electricity. Questions about nuclear power seemed distant.

Only an extraordinarily perceptive public, undistracted by other crises, might have forced the issue in the late sixties as it is doing today. Only a president and Congress sensitive to the abuses wrought by the AEC and the Joint Committee could have sought to correct them and bring the nation's attention to the nuclear power dilemma. The nation's attention, however, was focused elsewhere. Neither the public nor its elected officials were yet able to absorb the importance of dealing with the unresolved nuclear questions.

The late sixties was the last time that a decision to delay the expansion of nuclear power could have been implemented without causing substantial economic disruption. During the early seventies nuclear power commitments began to mushroom. In 1973 planned construction of nuclear power plants exceeded that planned for conventional power. The nation was becoming financially beholden as it had been psychologically and politically to the belief that nuclear power was inevitable.

Yet as more nuclear plants were built, more citizens became apprehensive. The Congress began to exhibit greater sensitivity as well, and in 1975 the AEC was abolished. The newly created Nuclear Regulatory Commission was charged exclusively with regulation, while promotion has become the responsibility of the Department of Energy.

The Joint Committee lost its designation as the sole overseer of nuclear affairs in Congress, and several congressional committees began to exercise interest in the subject. The structural aberrations created thirty years earlier were finally corrected, but their legacies include the uncertainties about nuclear power that continue to infuse the nation.

The inadequacy of congressional oversight was graphically revealed by Senator Abraham Ribicoff during congressional hearings in 1975. Speaking before the most enthusiastic nuclear supporters in Congress, and without refutation from a single member, Senator Ribicoff could say: "The fact is that no member of Congress today knows for sure what the safety and safeguards picture really is. That's the problem. We don't have enough information."[14]

No member of Congress, no member of the government body responsible for establishing laws to protect the public, was adequately informed about this potential threat to the well-being of the nation and its people. There were then 55 nuclear plants in operation and plans for 145 more by 1985.

THE LEGACY OF DISTORTED INSTITUTIONS

In considering the dilemma, I interviewed dozens of scientists, administrators, and others associated with nuclear development.[15] Was there ever serious discussion at any level of government about whether nuclear power might simply be too dangerous? None of the former or present members of the AEC, the NRC, the Joint Committee, or atomic scientists with whom I spoke could recall any debates about the overall commitment.

"I don't think there ever was much of a debate because those of us closest to the field never had any question but that it could be done safely," says a radiation physicist associated with nuclear development since the forties. Now a member of the government's advisory committee on reactor safeguards, he requests anonymity. "I guess that troubles the public, or certain segments of the public. They say, 'well these dumb people, they're promoting nuclear power, this hazard.' And yet we just never considered it in this way."

Have recent events, particularly the near-disaster at the Three Mile Island nuclear power plant, caused doubts?

"Not really," the anonymous physicist responds. "Look, we never told the people that accidents would never happen. We said that minor accidents would occur rather frequently and major ones less frequently. We never said that Three Mile Island events would not occur. We said the frequency, we hoped, would be rather low."

Like other proponents, he is convinced that nuclear power is reasonably safe and feasible. In the light of Three Mile Island and other unresolved questions, one may fairly wonder about the ingredients of such a conviction: How much derives from objective evidence, how much from the human need for justifying a thirty-year personal and professional commitment?

Social scientists have long understood the human tendency to retain deep-seated convictions in the face of contradictory evidence.[16] Old truths are not easily discarded.

Robert Pollard, then, is a curiosity. Hero to some, troublemaker to others, he resigned from the Nuclear Regulatory Commission in 1976, despairing about the nuclear power program. Pollard had been an engineer and project manager with the AEC and NRC for six years and before that was an instructor in the Navy's nuclear power program. He was committed to nuclear power and was highly competent. His performance appraisals are filled with superlatives about his expertise, his conscientiousness, and his managerial ability.[17] He seemed destined for higher levels.

Pollard cannot pinpoint the moment he decided to resign. He had been raising questions about safety violations with his superiors for years. Rarely were his complaints confronted directly. He was urged to go slowly, to modify his complaints, to drop them. He became increasingly uncomfortable.

"It was cumulative," he said. "To see an agency out there lying to the public, telling them that our plants are safe, telling the public that our regulations are being enforced; yet seeing on the inside how plants are being licensed that didn't meet the regulations, knowing the whole list of safety problems which show that regulations were inadequate. I just finally couldn't do it."

Pollard, now in his early forties, works with the Union of Concerned Scientists, an organization that is skeptical about the wisdom of pursuing nuclear power. He is serious and articulate, and he is saddened that people may not be taking his message seriously. As an engineer Pollard still finds nuclear technology esthetically appealing. But he has come to believe it can never be worth the risk.

"When I start hearing the industries cry about the benefits versus the risks, I don't know what they mean. The chance of having people exposed to an accident that never ends, that goes on from generation to generation; the fear of watching your children grow up and trying to decide when they should go to the doctor to have their thyroid operated on; the decision whether to have children for fear of having a deformed child—what benefits can be balanced against this? The comparison of how many people are killed by airplane crashes and cars—as bad as that is, at least the damn thing when it's over, it's over.

"Essentially you have to ask yourself if you're going to be in favor of nuclear power, how often is it acceptable to lose part of a state? If your answer is never, then it means you can't have nuclear power. Because you can't preclude a catastrophe."

If our political institutions had not been altered to accommodate nuclear affairs, would developments have been different?

"I really don't know if we could have done anything differently in the early stages," Pollard speculates. "When you start off with the idea of nuclear power, it sounds great. I can remember how thrilled I was in the nuclear-powered submarine program.

"But we should have been more sensitive as adverse information came to light. The problem has been to stop the momentum, to recognize you're going down the wrong road and do something about it."

As is evident, our political system was unprepared (and unwilling) to check the momentum. The institutional bulwarks had been stripped by the government's intense desire to have nuclear power seemingly at any cost.

From this and from the rest of our nuclear history, four lessons have become apparent. First, a potentially hazardous scientific or technological enterprise, no matter how alluring, should be pretested before a national commitment is made to its expansion. It is not satisfactory, for example, to be told that disposal of nuclear waste will present no problem. The technology should be demonstrated in advance. Despite earlier assurances, we still have no safe, permanent repositories, and there is question whether we ever will.

Second, as far as possible, the political and scientific aspects of an issue should be separately understood and separately evaluated. "It doesn't matter which valve fails at Three Mile Island," says Pollard. "The point is not that a valve failed. The point is what led up to the valve failing." By understanding this difference, we can understand that risks in nuclear plants arise fundamentally not from sticky valves but from political choice.

It follows that, third, we must be wary of tampering with our political institutions in order to expedite policy. To accommodate nuclear development we removed institutional safeguards. We learned that it was easier to strip the safeguards than to undo damage left by the stripping. A nuclear reactor engineer recently reported, for example, that steel vessels that house the radioactive core in many plants unexpectedly have become brittle and are on the verge of cracking and causing meltdowns.[18] Similarly, the Nuclear Regulatory Commission acknowledged in 1982 that as many as half the existing nuclear plants were "generally deficient" in their ability to respond to accidents.[19]

Finally, our nuclear history has been a reaffirmation of the need to expose issues fully before policies are formulated. The development of the atomic bomb was shrouded in secrecy. Secrecy became a habit. Insofar as nuclear power has been a nonmilitary enterprise it should have received the scrutiny applied to any other public issue.

If we had not been mesmerized by the nuclear mammon, if nuclear problems had from the beginning been publicly discussed, perhaps we would have developed nuclear power more slowly, perhaps not at all. As it stands, our nuclear power policies are largely a product of stifled democratic practices.

NUCLEAR WASTE

Perhaps the most awesome legacy of our nuclear program has been the continuing accumulation of nuclear waste. Some waste remains dangerously radioactive for tens of thousands of years. Despite assurances by experts that creating permanent waste repositories presents no problem, there is still no commonly accepted technique.

During interviews, some experts said that burial in geologically stable rock formations would be best. Others urged burial in salt mines, under the ocean floor, or rocketing waste into space. Still others regard placement beyond retrievability to be a mistake in case the permanent repositories prove fallible.[20]

The confusion arising from so many "best" solutions and the self-deception engaged in by official agencies for thirty years offer little comfort about the future. The unfortunate history of the nuclear waste issue is illustrated in the following chronology of statements and incidents. It is testimony to the inadequate governance by the political system of the nuclear waste problem born in large measure of wishful thinking.

> An AEC-supported project . . . is investigating the possible hazards of releasing radio-isotope waste into public sewers." (Atomic Energy Commission, *Semi-Annual Report,* January–June, 1951)

> Waste problems have proved completely manageable in the operation of the Commission and of its predecessor wartime agency, the Manhattan Engineering District, Corps of Engineers, U.S. Army. (Atomic Energy Commission, *Annual Report,* January–December 1959)

> A long-term study of low-level radioactive waste disposal operations at three Pacific Ocean sites . . . has shown that no radioactive hazard has been created, and that under present procedures an additional 40 sites could be established with safety. (Atomic Energy Commission, *Annual Report,* January–December 1962)

> Waste management has advanced more than is commonly believed and the problem nowadays is not so much concerned with the increasing quantities of waste as with the most economical methods of treatment and disposal. (International Atomic Energy Agency, *Basic Factors for the Treatment and Disposal of Radioactive Wastes,* 1967)

> The AEC chemical processing facilities at Hanford, Savannah River, and the National Reactor Testing Station in Idaho, concentrate and store radioactive waste material in large underground tanks or bins. . . . Tank storage has proved both safe and practical. (Atomic Energy Commission, *Annual Report,* January–December 1969)

Two underground storage tanks at Hanford developed leaks during 1974 bringing to 18 the number of Hanford tanks which have leaked and been removed from liquid storage service. At Savannah River, two double-shell tanks developed small cracks during the year, bringing the total to eight. . . . (Atomic Energy Commission, *Annual Report,* January–December 1974)

Reasonable progress toward the development of permanent disposal facilities is presently being accomplished. (Nuclear Regulatory Commission statement, November 1976)

It will, of course, be impossible to "demonstrate" with a high assurance of validity the capability of [a] repository to contain high level waste over the period for which it constitutes a potential radiological hazard. (Report prepared for the Energy Research and Development Administration, 1976)

Gerald M. Hansler, the Environmental Protection Agency regional administrator, noted that radioactive material already had contaminated a nearby stream [at the West Valley, New York repository]. "If the goal of low-level nuclear-waste disposal is 100 percent retention of the waste for the duration of its hazardous lifetime (300 to 1,000 years), then in 14 years West Valley has failed," he said. (*The New York Times,* February 10, 1977)

Since 1957, when a committee of the National Academy of Sciences first proposed the burial of [radioactive] wastes in deep geologically stable rock formations, a substantial body of evidence has accumulated pointing to the technical feasibility, economic practicality and comparative safety of this approach. (Dr. Bernard Cohen, "The Disposal of Radioactive Wastes from Fission Reactors," *Scientific American,* vol. 236, no. 6, June 1977)

A nuclear waste burial ground at Maxey Flats, Kentucky leaked radioactive waste into the groundwater. (*The New York Times,* April 16, 1978)

We are surprised and dismayed to discover how few relevant data are available on most of the candidate rock types even 30 years after wastes began to accumulate from weapons development. These rocks include granite types, basalts, and shales. Furthermore, we are only just now learning about the problem of water in salt beds, and the need for careful measurements of water in [salt] domes." (Report by panel of distinguished earth scientists requested by the Environmental Protection Agency, 1978, cited in *Science,* vol. 200, no. 4346, June 9, 1978)

Given the current state of our knowledge, the uncertainties associated with hot wastes that interact chemically and mechanically with the rock and fluid system appear very high. . . . Construction of a repository and emplacement of waste will initiate complex processes that cannot, at present, be predicted with certainty. (Report published by the U.S. Geological Survey, *Geologic Disposal of High-Level Radioactive Wastes—Earth Science Perspectives,* 1978)

Government officials said today that more old radioactive waste dumps of the kind discovered this month in Denver were probably scattered around the country. But the officials said that nobody knew how many such sites there were, where they were situated or how much of a hazard they presented to public health. (*The New York Times,* February 25, 1979)

A special committee of the Federal Government reported to President Carter today that the safety of disposing of high-level radioactive wastes in underground repositories could be determined only after specific investigations at particular sites. The committee's assessment conflicted with the broad assertions of Federal officials over the years about the relative ease of handling radioactive wastes created by nuclear power reactors and the manufacture of nuclear weapons. (*The New York Times,* March 14, 1979)

I really think that technically it's a grossly exaggerated problem. . . . To sequester something for 800 years in an area where there is little earthquake activity just doesn't seem to me to be all that big a deal. (Dr. Alvin Weinberg, former director of Oak Ridge National Laboratory, interview, May 1, 1979)

The Carter Administration is considering buying a small, uninhabited island in the South Pacific for storage of nuclear waste from Asian reactors, Government officials said Friday. The officials said that Administration experts were examining whether Palmyra, in the Line Islands and 1,110 miles southwest of Hawaii, would be a good place to store waste." (*The New York Times,* August 19, 1979)

"We do not feel we are being unpatriotic or un-American," said Ainsley Fullard-Leo, who with two brothers owns the 1,400-acre atoll. . . . "The laugh was a bitter one for us when we read in the papers that the Government wanted to buy Palmyra and turn it into a nuclear dump. . . . We have all agreed that no matter the price, we will never sell Palmyra to anyone wanting to store nuclear wastes there." (*The New York Times,* August 23, 1979)

> There's no hurry on nuclear waste. . . . Though the task [of disposal] sounds formidable, there is reason for optimism. (Editorial, *The New York Times,* December 27, 1979)

> President Carter will soon announce a national policy on nuclear waste that will postpone the opening of any permanent disposal site until about 1992." (Washington Post News Service, reported in *The Record* [Hackensack, N.J.], January 10, 1980.)

> Just out is a study for the government on waste disposal, entitled "Archaeological Data as a Basis for Repository Marker Design." . . . The conclusion: Waste sites should be marked by a series of giant stones, similar to the ancient megaliths of Stonehenge in southern Britain. . . . The markers [would] be inscribed on three sides with a warning: "Danger. Radioactive waste. Do not dig deeply here." . . . This warning would then be repeated in the six languages of the United Nations: French, Arabic, Spanish, Russian and Chinese. [The] presumption is that someone will be speaking one of these tongues 10,000 years from now. But if not, [the study] has also developed a warning symbol for the marker patterned after the international driving signs. The symbol shows a human digging and has a line drawn across it. (*The New York Times,* November 25, 1982)

Despite the insistence by some experts that creating a safe permanent repository for nuclear waste poses no serious problem, this thirty-year-old plea evidently has worn thin. A majority of scientists in our survey was skeptical. Sixty-six percent consider it a major technological problem and only 28 percent think it is not, as shown in Table 8.1. Ralph Lapp, a respected physicist who supports nuclear power expansion, characterizes skeptics as "people who have no technical competence whatsoever, who have no degrees in geology or physics."[21] In fact the eighty physicists and twenty-eight geologists who responded to the question on nuclear waste repositories divided much like the overall sample of scientists. Only 33 percent of the physicists agreed that creating safe permanent repositories presented no major technological problems, while 66 percent disagreed. Skepticism among geologists was even more striking. Seventy-five percent disagreed with the proposition, 57 percent strongly.[22]

Perhaps the optimism about nuclear waste will eventually prove correct. At the moment no one can know. Proliferation of nuclear waste on the hope and expectation that adequate

Table 8.1 Responses of Scientists to Question 21 of the Survey (by Percent)

As I understand it, creating a safe permanent repository for nuclear waste is not a major technological problem.

	A	B	
	All respondents	Physicists	Geologists
Strongly agree	5	4	7
Agree	23	29	18
Disagree	38	43	18
Strongly disagree	28	23	57
Can't say/ no answer	7	3	0
Total	101 (632)	102 (80)	100 (28)

repositories will be developed has been a political judgment. The prudent course would have been to demonstrate a successful technology before proliferation. But the failure of the government to limit nuclear expansion has confronted us with a seemingly irreversible problem.

Recombinant DNA—A Better Way

A recent example of a more successful political approach to an issue involving science policy suggests how nuclear power might have been handled, and indeed still might be.

In the early seventies, conditions seemed ripe for a biological rerun of the atomic experience. A technology was discovered in which genes could be transferred from one organism to another. Experimentation with recombinant DNA, as it was called, was touted as the key to curing cancer, heart disease, immunological malfunctions, and other illnesses.

David Baltimore, a Nobel prize winner for his work with tumor viruses, spoke for many biologists. He was convinced that recombinant DNA research would lead to elimination of "the presently unconquered diseases that afflict us."[23] Baltimore and others urged full federal support to pursue the new millennium. But this time there would be no headlong plunge.

A current of skepticism had arisen among several molecular biologists about the safety of recombinant DNA research. Some felt that genetic tampering might produce dangerous organisms or otherwise harm future generations. A series of scientific conferences was held during the mid-seventies, and a moratorium on DNA research was instituted.

The issues were reported by the news media and debated in city councils and before congressional committees. Ultimately the National Institutes of Health issued guidelines that were restrictive in proportion to the perceived danger of the experiment. Some types of experiments were prohibited entirely.[24]

The manner of developing recombinant DNA policies stood in sharp contrast to that of nuclear power. Nowhere is this better demonstrated than by the effort to disseminate broadly the points at issue. The proceedings of a conference on recombinant DNA sponsored by the National Academy of Sciences, for example, were published as "an attempt to help the layman and the scientist understand those facts that are already established, those facts that are in dispute, and the practical and moral considerations that must be part of any attempt to make final policy." Readers and observers were urged to examine the arguments of the DNA protagonists "at leisure," and "hopefully, this will allow a wider constituency to inform themselves and clarify the issues in this vital area of scientific development."[25] The spirit of this approach could hardly have been more at odds with the one prevailing during the development of our nuclear policies.

When Roy Curtiss, a microbiologist at the University of Alabama, recalls the DNA debate, he imparts a sense of satisfaction. "I think that in the years to come the scientific community will look back on our effort and regard it as a good thing. I think we'll feel good about our involvement."

At first Curtiss had opposed recombinant DNA research. As the issues were aired and as he considered new experimental evidence, he softened his opposition. He helped draft the NIH guidelines. Curtiss had worked for ten years at the AEC's Oak Ridge National Laboratory and was a close observer of nuclear as well as DNA developments. The political system has acted far more responsibly in the DNA than nuclear areas, he believes. The most poignant difference between the two, he says, has been the secrecy of the nuclear environment compared with the openness of DNA developments.

When the time came to confront the recombinant DNA issue, many scientists acted in response to the bruises suffered from the nuclear experience. Some were appalled at the scientific misinformation raised in several public debates about DNA. But compared to the nuclear issue, DNA stands as a celebration of the American political system.[26]

At each step recombinant DNA has been handled as nuclear power was not: comprehensive pretesting before commitments to expand the technology, recognition of the distinction between political and technological arguments, employment of traditional political institutions, open debate.

Above all, the nuclear lesson should stand as a warning about other scientific and technological activities that may pose irreversible threats. Though none now seems as formidable as nuclear waste proliferation, new activities or products like pesticides, building materials, food additives, cosmetics, and other intended benefits to mankind have proved unexpectedly hazardous. In consequence, the thought of restraining scientific activity, while abhorrent to some, must be understood in the context of changed conditions.

Notes

1. Transcript of meeting of Nuclear Regulatory Commission reproduced in *The New York Times*, April 14, 1979, p. 9.

2. Interview, April 15, 1979, and also David E. Lilienthal, *Change, Hope, and the Bomb* (Princeton, N.J.: Princeton University Press, 1963), pp. 18–19.

3. Harold P. Green and Alan Rosenthal, *Government of the Atom* (New York: Atherton Press, 1963).

4. Interview, April 26, 1979.

5. U.S. Congress. Joint Hearing before the Subcommittee on Oversight and Investigations of the Committee on Interstate and Foreign Commerce, House of Representatives, and the Health and Scientific Research Subcommittee of the Labor and Human Resources Committee and the Committee on the Judiciary, United States Senate, *Health Effects of Low-Level Radiation, Volume I*, April 19, 1979 (Washington, D.C.: Government Printing Office, 1979), p. 180.

6. In 1956, the sheep owners sought compensation from the government but lost their case in a trial before Federal Judge A. Sherman Christensen. At the request of the sheep owners, Judge Christensen, now seventy-seven years old, reviewed the case in 1982 and ordered a new trial in view of the apparent fraud by government lawyers and scientists. *Science*, vol. 218, no. 4572 (November 5, 1982), pp. 545–47.

7. *The New York Times*, November 4, 1956, p. 62.

8. John G. Fuller, *We Almost Lost Detroit* (New York: Readers Digest Press, 1975), esp. chapters 3, 13–14.

9. Richard Curtis and Elizabeth Hogan, *Perils of the Peaceful Atom* (New York: Doubleday and Co., 1969), pp. 194–96.

10. See typical comments by Senator Wayne Morse. U.S. Congress. Hearings before the Joint Committee on Atomic Energy, *Proposed Legislation for Accelerating Civilian Reactor Program*, May 23, 24, 25, 28, 29, 1956 (Washington, D.C.: Government Printing Office, 1956), pp. 125–27.

11. Interview, April 15, 1979.

12. Ralph Nader and John Abbotts, *The Menace of Atomic Energy* (New York: W. W. Norton and Co., 1977), p. 70.

13. Interview, May 1, 1979.

14. U.S. Congress. Joint Hearing before the Joint Committee on Atomic Energy, Congress of the United States, and the Committee on Government Operations of the United States Senate, *Nuclear Regulatory Commission Action Requiring Safety Inspection which Resulted in Shutdown of Certain Nuclear Power Plants*, February 5, 1975 (Washington, D.C.: Government Printing Office, 1975), p. 4.

15. Unless otherwise noted, comments attributed to individuals are based on interviews conducted in 1979 and 1980.

16. A classic study of political behavior found many voters convinced that issues they supported were also supported by their favored presidential candidates, even when the candidates in fact opposed the issues. See Bernard R. Berelson, Paul F. Lazarsfeld, and William N. McPhee, *Voting, A Study of Opinion Formation in a Presidential Campaign* (Chicago: University of Chicago Press, 1966), pp. 215–33.

17. U.S. Congress. Joint Committee on Atomic Energy. *Investigation of Charges Relating to Nuclear Reactor Safety*, vol. 1, February 18, 23, and 24, and March 2 and 4, 1976 (Washington, D.C.: Government Printing Office, 1976), pp. 125–27.

18. Demetrios L. Basdekas, "The Risk of a Meltdown," *The New York Times*, March 29, 1982, p. A-19.

19. *The New York Times*, June 23, 1982, p. A-16.

20. The variety of opinions was expressed during interviews with nuclear scientists, but is also found in literature dealing with nuclear waste. Nader and Abbotts, pp. 148–58; Philip M. Boffey, *The Brain Bank of America* (New York: McGraw-Hill Book Co., 1975), pp. 90–111; Sheldon Novick, *The Electric War, The Fight over Nuclear Power* (San Francisco: Sierra Club Books, 1976), pp. 177–80.

21. Transcript on "Nuclear Waste," *The MacNeil-Lehrer Report*. New York, WNET, air date July 27, 1978, p. 7.

22. A series of national surveys indicate that public concern about waste disposal has been mounting for several years, and was greater than about any other issue associated with nuclear power (though the surveys were taken prior to the accident at Three Mile Island). Barbara D. Melber, Stanley M. Nealey, Joy Hammersla, William L. Rankin, *Nuclear Power and the Public: Analysis of Collected Survey Research* (Seattle, Wash.: Battelle Memorial Institute, November, 1977), pp. 170–77.

23. Nicholas Wade, *The Ultimate Experiment, Man-Made Evolution* (New York: Walker and Co., 1977), p. 115.

24. The guidelines, and summaries of several conferences held during the seventies, are reproduced in Clifford Grobstein, *A Double Image of the Double Helix, The Recombinant DNA Debate* (San Francisco: W. H. Freeman and Co., 1979), pp. 113–71.

25. See *Research with Recombinant DNA, An Academy Forum, March 7–9, 1977* (Washington, D.C.: National Academy of Sciences, 1977), p. 3.

26. A few molecular biologists reconfirmed their opposition to recombinant DNA research including Robert Sinsheimer, Erwin Chargaff, and Liebe Cavalieri, during interviews in 1979. But most who had been skeptical earlier felt that existing guidelines were adequate to insure safety. The political and scientific debates are recounted in several books including those by Wade and Grobstein. Also see John Lear, *Recombinant DNA, The Untold Story* (New York: Crown Publishers, 1978); and Michael Rogers, *Biohazard* (New York: Alfred A. Knopf, 1977).

9

External Interference: The Scientist's Quandary and Government's Responsibility

As alluded to previously, changes in the nature of scientific activity have introduced new challenges. Whether American political institutions can adequately address these challenges is uncertain. Before World War II the consequences of scientific or technological activity, however dangerous, were reversible. Even purposefully destructive applications were self-limiting. When a military battle was over, destruction ceased. Although the seventeenth-century church, Stalin's Soviet Union, or Hitler's Germany could impose scientific truth, the imposition would cease upon demise of the regime. Scientific verities and scientific activity were reversible.

Nuclear science introduced a qualitative change that has bred confusion about the proper relationship between political authority and scientific inquiry. The question was heightened during the seventies with the controversy about recombinant DNA research and restraining scientific inquiry in general. By the end of the decade, scientists and nonscientists were wrestling with questions about the legitimacy of governmental interference with scientific activity. The traditional belief "that no limit can or should be set upon scientific inquiry," as Gerard Piel continued to advocate, was under siege.[1]

A wide range of views was expressed at several conferences, and the elusiveness of solutions became evident at one session after another. In the concluding essay presented at a conference on "Scientific Expertise and the Public," Dorothy Nelkin and Judith Swazey raised questions central to the dilemma:

> Is there some research so threatening to the basic values of certain groups or so potentially risky to human subjects that it should not be done at all? Who are, or ought to be, the "experts" in decisions about the nature and governance of research? Is the traditional freedom of scientists to define and control their own research still reasonable given the expanded possibilities of modern science in areas such as human biology and behavior?[2]

Gerald Holton, in an epilogue to a symposium on the "Limits of Scientific Inquiry," acknowledged that "we have only begun to struggle with such problems."[3]

At a symposium sponsored by the American Association for the Advancement of Science on the "Regulation of Scientific Inquiry," several contributors made specific proposals about restraining scientific activity. Their positions revealed the enormous disparity among scholars and scientists on the issue. Keith Wulff concluded that "in general, scientists should regulate themselves." He based his conviction on the questionable assumption that "the people most critical of science have so far been the scientists themselves."[4] His position seems grounded in the belief that scientists are better endowed (by intelligence? by knowledge? by training?) to restrain themselves than are other professionals: "Self-regulation has not worked as well as one might hope in the medical and legal professions; however, I do not think that these professions are analogous to the scientific community."[5]

The argument against all external regulation was dramatically intoned in the AAAS symposium by Piel, Wulff, and others against the specter of the Galilean, Lysenko, and Nazi episodes. It should be clear from this study that the frequent practice of comparing the historic cases with governmental interference in the United States is invalid and a liability to constructive analysis. This was inferentially recognized by some at the AAAS symposium who argued against scientific arrogance. Andre Hellegers sensed that many scientists had exchanged the authority of earlier intolerant systems for that of an intolerant science:

> I sometimes think we are assuming the posture of the medieval clergy, who would insist that they should not be scrutinized because their product was eternal life and truth (albeit in the hereafter). Similarly, some insist that we are the custodians of eternal life and verities (in the here and now), so that we should also be immune to inquiry, while we insist on our own freedom of inquiry.[6]

While Hellegers would restrain scientific and technological activity known to be dangerous, Barry Casper goes further. He would restrain research leading to *knowledge* that "might be dangerous."[7] He offers as an example research to enrich uranium by means of lasers. Such efforts could lead to a cheap, simple technology that would make weapons-grade nuclear fuel available to nations that cannot now obtain it. Casper would prohibit such research.[8]

The variety of positions at this and other symposia ranged from favoring substantial restraints, as did Casper, to favoring none, as did Piel. But the number of contributors at all the conferences devoted to the issue totaled only a few dozen. In seeking the views of a larger sample, several questions relative to the issue were asked in this study's survey of scientists.

Government and Science—Perceptions by Scientists

SPECIFIC ACTIVITIES

Based on the perceptions among survey respondents about specific scientific or technological activities, it is evident that uncertainty cuts through the scientific community. With the exception of nuclear waste proliferation, which 75 percent of the respondents perceive as very or moderately risky, there is no consensus about which activities are particularly dangerous to the public. (See Table 9.1.)

While 44 percent think recombinant DNA research is at least moderately risky, 47 percent consider it slightly or not at all risky. Although the division on the risks attendant to nuclear power plants is more pronounced—40 percent see some risk compared to 57 percent who see little or no risk—it is not decisive. A similar division arises between the 58 percent who find existing levels of environmental pollution moderately or very risky and the 39 percent who do not.

Table 9.1 Responses of Scientists to Question 19 of the Survey (by Percent)

Certain scientific or technological activity may create risks to public safety. How do you feel about the following:

	Very risky	Moderately risky	Slightly risky	Not at all risky	Can't say/ no answer	Total
Recombinant DNA research	13	31	40	7	10	101 (632)
Nuclear waste proliferation	41	34	21	2	3	101 (632)
Nuclear power plants	11	29	49	8	3	100 (632)
Existing levels of environmental pollution	19	39	36	3	3	100 (632)

Though invited to enter on the questionnaire other scientific or technological activities they deem risky, most scientists demurred. Even the 20 percent who did offered a striking variety among their responses. The most frequently cited were activities relating to military (including nuclear) armament and those involving chemicals-drugs-food additives. Each of these two categories drew mention from only 3 percent of the total number of respondents. Activities falling in nineteen other categories were cited, though none by more than 2 percent. They included such diverse concerns as research in biological and chemical warfare, exhaustion of our nonrenewable resources, chemical waste, increased atmospheric carbon dioxide, acid rain, cloning, behavior modification, climate control, solar power stations, x-rays, disease research, and electromagnetic waves.

Trying to determine which scientific or technological activities pose the most serious risks to the public by polling eminent scientists seems fruitless. Their perceptions are too disparate.

GOVERNMENTAL INTERFERENCE

On the question of governmental restraint of scientific activity, the scientists divide almost evenly in principle. In Table 9.2,

Table 9.2 Responses of Scientists to Question 16 of the Survey (by Percent)

The government should never interfere with scientific activity that genuinely seeks to increase knowledge.

Strongly agree	Agree	Disagree	Strongly disagree	Can't say/ no answer	Total
17	28	43	5	7	100 (632)

Table 9.3 Responses of Scientists to Question 26 of the Survey (by Percent)

Scientific activity that may be hazardous should be:

Regulated by the government	32
Regulated by designated scientific experts	45
Regulated by combination of government and designated scientific experts*	15
Not formally regulated beyond the good sense of the experimenter	2
Other/can't say/no answer	6
Total	100 (632)

*Though not one of the choices delineated on the questionnaire, this category is listed here because of the substantial number of write-ins under "other."

45 percent feel that the government should not interfere with scientific activity that genuinely seeks to increase knowledge, while 48 percent disagree. Even when the possibility of hazard to the public exists, more scientists feel that regulation should rest with scientific experts than with the government. (See Table 9.3.)

The scientists' aversion to governmental restraint is exemplified in their uncertainty about the degree to which the government has been *overrestrictive* on nuclear developments. As shown in Table 9.4, 32 percent believe that nuclear power could have been expanded more safely and effectively if it had not been impeded by external politics, and 25 percent were not sure. Despite the history of weak governmental oversight, of often blind encouragement of nuclear develop-

Table 9.4 Responses of Scientists to Question 22 of the Survey (by Percent)

If the work of nuclear scientists and technicians had not been impeded by politics external to their discipline, nuclear power could have been expanded more effectively and safely.

Strongly agree	Agree	Disagree	Strongly disagree	Can't say/ no answer	Total
8	24	31	13	25	101 (632)

Table 9.5 Responses of Scientists to Question 20 of the Survey (by Percent)

Restrictions imposed by the government on recombinant DNA research constitute unjustified interference with discovery of scientific truth.

Strongly agree	Agree	Disagree	Strongly disagree	Can't say/ no answer	Total
5	21	53	6	15	100 (632)

ment, fewer than half the scientists in our survey seemed to acknowledge this. Only 44 percent disagree with the assertion that the government effectively held back expansion of nuclear power.

On the issue of recombinant DNA research, just 26 percent agree that governmental restrictions constitute unjustified interference. Fifty-nine percent accept the need for some governmental regulation. (See Table 9.5.)

When comments were solicited about governmental interference with scientific activity, few respondents rejected a role for government entirely. Most recognized that the ultimate protection of the citizenry lies in its political system. They inferentially accepted the axiom that the public welfare cannot be left to any private interest, but rather to the only mechanism that can speak for the total polity, its government.

To the question "How does one decide when governmental interference in scientific activity is legitimate or not?" hardly any respondents explicitly denounced *all* political intervention. Only 5 percent offered comments like:

> I can think of no examples where governmental interference has improved the situation.

> Unfortunately the government's response is frequently disastrous rather than beneficial.

> Very simple. No governmental interference in scientific activity is either legitimate or safe. Any governmental authority including granting of research support is illegitimate. This has allowed dreadful fiascos in modern-day health sciences.

About 17 percent made clear in their comments that scientists alone should decide when interference was legitimate. Typical comments from this group included:

> The scientific competence of a politically balanced spectrum of scientists in the activity can decide. All else is not legitimate.

> A consensus of leading scientists in the particular field should be ascertained through a poll.

> Decisions or restrictions should be made by peers in science, not selected by or subjected to political pressures.

> A scientific evaluation (not political) of all existing data should be carried out in order to assess risk-benefit.

> It is not easy. You will have to depend on informed scientific opinion. These experts in turn must be people with a breadth wider than that imposed by their scientific specialty.

> By discussions with scientists most closely involved in the activities in question, and by application of common sense.

> A tough question. In general, the scientific community itself is probably best capable of raising the question and deciding upon the necessity for governmental interference.

> In all fairness it is very difficult. I believe that scientists should monitor such supposed interference, raise questions within the scientific community and bring the view of the discipline to public attention.

About 20 percent of the respondents unequivocally cited the government or the public as the ultimate legitimate authority in deciding when to interfere with scientific activity. Typically:

> The government (public) has the right to spend its money as it chooses and to control those nonpolitical activities which present a potential danger to the public.
>
> It is legitimate for government to interfere when the public welfare is not adequately being considered by the scientific community.
>
> If government is paying, government must decide if activity reflects a need of government. If government is not paying, it must decide if activity may result in harm to the public.
>
> When scientific activity is expected to have significant impact on the well-being of the entire public, or the future of society, it becomes a matter of the society, and thus the government.
>
> When one perceives that the activity may endanger the health and welfare of individuals within the society, then one must resort to government regulation. Only the government is responsible to the people; not the scientists involved.
>
> In a representative form of government such as ours, the will of the public should decide, as expressed through the actions of elected politicians.
>
> If there is any question whether present or future life will be adversely affected, governmental interference is legitimate until the question is resolved.
>
> Scientists, being human, cannot be trusted to police themselves. Research that seriously endangers life, human or otherwise, must be policed.

Nineteen percent of the respondents did not comment. But in addition to the 20 percent who unequivocally supported the right of government to interfere, another 41 percent implicitly acknowledge the legitimacy of governmental interference at some point. The answers were often unfocused but understanding of its necessity at times. The following excerpts reveal a mix of frustration, oversimplification, perceptiveness, yet recognition of the government's ultimate responsibility:

> Whether or not it involves (real) public safety; when based on facts and not mass hysteria; when not obviously for political gain.

Through open debate such as is now taking place on nuclear power and other issues. This is hampered somewhat by the nature of media coverage and dispersion of information which seems to find commercial value in excitement to the detriment of factual information based on intrinsic importance of the issues.

Tough question, and basically a political one. Our political system requires limited consensus (majority), so in last analysis this question must be answered by a general referendum based on fairly presented points of view.

This is a political question. The Supreme Court decides such matters.

This is a hard question. It would be helpful to have a series of cases and to rate each case rather than give broad principles.

It's hard to say because when you come down to it, even governmental grants (and who gets them) are forms of interference; and yet they are considered legitimate and welcomed. Also it is dependent on the time frame—what was good and supported by government in the 1950s might look very bad in the 1970s and 80s, thus emphasizing the difficulty of the question.

As a Christian and a scientist I believe our human wisdom is woefully inadequate to make such decisions. I would have us refer to absolute authority: God's law. Find how it applies to the scientific activity and/or to the regulations, and base the decision on my understanding of what God's word says to the situation.

Common sense.

By objective analysis of the situation.

The same way porcupines make love.

Legitimacy depends so much on one's point of view. I'm not sure if a valid decision can be made in many cases. It's a matter of judgment and debate.

There is no general rule applicable to all cases; common sense should prevail.

Good question!

Depends on who heads the government.

Difficult, but probably when it is to the benefit of the governed (but who decides that?).

There is no correct answer. If one could determine what is best for our overall ecosystem, that would be in the right direction. Knowledge, education, the political process, and the times all play a role.

It is not easy. But the public must be protected. Some scientists seem to be so interested in their research activities that they are unaware of hazard.

Deontologically.

When there is a criminal activity involved.

Very difficult when subtle, as in funding research deemed "worthy." In many instances, like human experimentation, the need is obvious, but in all cases it is a sad reflection on the nature of scientists and their need for surveillance.

One measure might be to ask who is objecting to interference. If objection is principally from the scientist himself, and from those who would profit financially, interference is probably justified.

The government should interfere only to protect the health and safety of the public. The problem is that a government can use this as an excuse for interference at any time, as witnessed by the many illegitimate actions in the name of "national security."

Purely subjective. One scientist's perversions are another's persuasions.

As perceived by scientists in this survey, the issue of governmental interference with scientific activity seems a virtually insoluble dilemma. Most responses reveal uncertainty, frustration, and simplistic proposals. But while a wide range of emphases are evident, the comments do bear common inferences.

First, although most scientists acknowledge the legitimacy of governmental interference with reluctance, they do recognize that it is necessary for some scientific activity. Second, although most do not specify what issues they feel warrant interference, those who do generally cite threats to public safety as due cause for such action. Third, though acknowledging necessity, scientists are not in agreement about how such decisions should be made, or who in particular should make them. Finally, in recognizing the complexity of the issue, many suggest that it would be impossible to apply a single standard to all cases.

The need for standards based on public safety should be beyond dispute. Potentially irreversible activities like radioactive waste proliferation, destruction of the ozone layer above the earth, or recombinant DNA research deserve particular concern. But should all scientific activity that appears dangerous be prohibited, as Hellegers suggests? Should prohibition be more comprehensive and include discovery of knowledge that might lead to dangerous consequences, as Casper urges? The answers can never be absolute, but a more enlightened method than now exists to determine which activities to limit should be welcomed.

Educating the Polity—Whistle-blowing and a Science Hearings Panel

Whistle-blowing, the exposure of wrong doing from within an organization, is a relatively recent phenomenon.[9] The scientific community's interest in whistle-blowing was highlighted in 1975 by a report of the American Association for the Advancement of Science entitled *Scientific Freedom and Responsibility.* The report applauded instances of scientists speaking out, such as the physicists who challenged the Atomic Energy Commission's radiation standards or the engineers who warned about dangers built into the San Francisco area rapid transit system. After the dissenters were rebuffed by their superiors, they spoke to the public and lost their jobs.[10] Their warnings ultimately proved valid, yet their experiences must dishearten others who might wish to speak out.

Although we have noted that most scientists are unlikely to oppose a politically mandated truth, we may wonder about their reactions to more threatening situations. How would they behave in the face of potentially catastrophic scientific activity? Our survey confirms an uneasiness that scientists feel about the issue.

The sample was asked what a scientist should do when he fears that the consequences of scientific activity with which he is familiar might be disastrous. Sixty-seven percent in Table 9.6 felt that he should raise his concerns in public. Only 28

Table 9.6 Responses of Scientists to Questions 23 and 24 of the Survey (by Percent)

	23. What should a scientist do when he fears that the consequences of scientific activity with which he is familiar might be disastrous? Should he:	24. What do you think most American scientists would do under the circumstances proposed in the previous question? Would they:
Engage in other activity and say nothing	0	10
Raise concerns in public	48	19
Raise concerns only within the scientific community	28	48
First raise concerns within scientific community, then in public*	19	7
Other/can't say/no answer	4	16
Total	99 (632)	100 (632)

*Though not one of the choices delineated on the questionnaire, this category is listed here because of the substantial number of write-ins under "other."

percent believed he should limit his concerns to expression within the scientific community. (Included among the former are several who urged that concerns first be discussed within the scientific community, but if the issue could not be resolved there, it should be brought before the public.) When the question was altered to ask what they thought most American scientists *would* do under these circumstances, only 26 percent believed that most would raise their concerns in public.

Some respondents inserted unflattering characterizations about their peers under the proposed circumstances. "Scientists have an ivory tower syndrome," said one. "Look what happened to Rachel Carson—she never got any recognition by U.S. scientists." Others cited financial motives: "They'll do whatever serves their economic interests." Or: "They'll engage in other activity and say nothing, unless it's lucrative."

Most respondents failed to comment on the question beyond registering their belief that their colleagues would not bring their concerns to the public. Whether based on economics, fear, or laziness they understood that reluctance to take a public stand was wrong. Many who did comment thought their colleagues' reluctance should not seem surprising. "Scientists are no more brave than the average man," wrote one respondent. "In fact the individual scientist may be somewhat more inhibited about taking risks because of his training." Another stated that "some people are very courageous, but most don't like to be thought odd by believing things that most of their friends and colleagues don't believe. The fact that scientists are involved does not distinguish this human trait from other people."

As the respondents infer, since scientists are surrounded by the same societal influences as nonscientists, they may be expected to react similarly. Substantial literature exists on the subject of social conformity. Weisband and Franck stress that efficiency and order are values in all organized, technocratic societies. "Team play," they point out, has come to override individual accountability in the American tradition.[11]

Others list concrete incentives to conform, such as job vulnerability and doubts that complaining in public would be effective in making changes.[12] Professional ambition and "social disapproval of tattling" also stand as powerful disincentives to whistle-blowing.[13] Nevertheless, if a reputable scientist suspects that certain scientific activities may lead to disaster, he has no higher obligation than to bring the matter out.

As a proposal for a better method to protect the public, the following model, which draws from lessons in this study, is conceived of in four stages. It is proposed as a series of structural principles, not a comprehensive formula. Its thrust is to promote exposure of the issues, education of the public, and informed decision-making by the government.

A first step would involve encouraging scientists to call the public's attention to their concerns about dangerous scientific activities. Scientists are the first to know about potential dangers from their work or that of their associates. While a large majority apparently would not bring their concerns to the public or government, most recognize that ultimate responsibility to protect society from danger lies with the government. The frustration that many feel about the paradox has been reflected in responses reported in this chapter. Whatever the reasons—lethargy, fear of ridicule or loss of research grants—a means of minimizing reluctance should be sought.

One method might involve creation of a Science Hearings Panel to receive reports of proposed or ongoing scientific activitiy deemed dangerous. The panel could be composed of several distinguished scientists and an equal number of nonscientists appointed by the President for fixed terms. Petitioners would initially submit a written document of concern, but might be invited to clarify their positions in person. As incentives, recognition and awards could be bestowed on those whose petitions to the panel prove valuable. If a petitioner preferred anonymity, this should be respected.[14] In any case, the priority would be to institutionalize a means to encourage scientists and others to express suspicions of danger. Whistle-blowing in the interest of public safety should be honored, at least by protecting the whistle-blower and his job.

Second, all inquiries or expressions of concern, except those obviously frivolous, should be published periodically, along with responses from experts in the field. Existing oversight agencies should be invited to comment as well. In areas involving research on humans, the appropriate commission for the protection of human subjects would be expected to make a statement. In questions about the use of certain chemicals, it would be the Food and Drug Administration, or perhaps the Environmental Protection Agency.

Government agencies appropriate to almost every area of scientific concern already exist. Their comments along with the petitioners' and those of recognized experts would receive exposure through publication by the Science Hearings Panel.

If the published responses do not resolve the questions about ethics or safety, a third step would be necessary. The Science Hearings Panel would convene a forum including experts and nonscientists to be participants and observers. The arguments would then be publicly debated and reported in the media.

The mechanics of a forum could employ an adversarial process approximating that for the proposed Science Court.[15] The forum, however, would differ fundamentally in purpose. Unlike the court, whose judges would ostensibly decide issues of fact apart from values, the forum would offer no verdict or decision. This is based on an implausible premise about the court—i.e., that facts and values can be neatly separated during an adversarial proceeding.[16] In fact, evidence presented to support any position is likely to be value-laden insofar as one advocate chooses it to enhance his argument. To presume that judges could ignore the advocates' values or their own when rendering a decision is questionable.

An example posed by proponents of the Science Court buttresses this contention. They suggest that the court could decide whether a particular nuclear power plant should be licensed or not based on facts alone.[17] The problem is almost self-evident. What facts? Do they include the attitude of nearby residents, that is, their values? Are the facts to be based on chances of an accident, which vary from one study to another? Who is to decide which set of statistics is correct?

This is not to say that elements of the court's proposed structure could not be useful—only its narrowly designated functions and purposes. An adversarial process including cross-examination by opposing advocates might best expose the issues of contention and could be employed at a forum. A forum should be an educational medium, not decision-making. Disposition of the issues should lie, finally, with the people's representatives, the government. Employment of a forum would create a far larger base of informed citizens and representatives to make decisions than has been usual in the past.

The proposed model resembles the sequence that evolved in consideration of the recombinant DNA issue. In the early seventies a few scientists were worried about the newly devised capability to recombine genetic material of different organisms. Their concerns were first reported in *Science* magazine and later in the mass media. Several publicized conferences were held during the mid-seventies at which supporters and critics of DNA research presented their views (though not in formal adversarial proceedings). The issues were debated within local and state governmental bodies, and advocates testified before congressional committees. Public exposure to the issues was broad.

The National Institutes of Health was assigned the task of formulating policy, and its proposed guidelines for recombinant DNA research received wide dissemination in the scientific community and the popular press. By the end of the seventies not all questions had been resolved, though a consensus on regulated research had been struck. From the start, the issues were considered and disposed of before the public.

The four stages in the recombinant DNA process ought to become the standard for all scientifically contentious issues with potential public impact. The concern initially raised by a few scientists about DNA research was welcome but uncommon. A Science Hearings Panel to which scientists and others may bring their concerns without fear of penalty would make the process easier.

Second, though *Science* reported the controversy and helped provide information later disseminated by the popular press, the development was fortuitous. A journal published by the Science Hearings Panel could reduce the uncertainty about exposure for other issues. Third, a formalized means to convene a forum would avoid uncertainties about ground rules and effectiveness that troubled conferences held on DNA policy. Finally, as with the DNA controversy, after broad public exposure the issue may better be understood and disposed of by governmental bodies, whether local or national, legislative or executive.

The suggested steps would require little innovation in political structure. A panel to act as a repository for scientific concerns, to publish responses, and to sponsor an occasional forum would constitute the only substantial innovation. The intention is not to organize a new system of judgment but to enhance the process of decision-making by the existing political apparatus. Scientific issues that affect the public should be decided no differently from nonscientific issues—before the people and their representatives.

We may anticipate increasing acceptance within the scientific community of the legitimacy of governmental restraint, as will be shown in the concluding chapter. Although many scientists remain suspicious of governmental interference, respondents to this study's survey demonstrated their concerns about hazardous scientific activities and the need to control them. However reluctantly, most scientists seem to recognize that the government ultimately is the only suitable body to impose restraint on scientific activities in the interest of the larger population.

Notes

1. Gerard Piel, "Scientific Research: Determining the Limits," in Keith M. Wulff, ed., *Regulation of Scientific Inquiry* (Boulder, Col.: Westview Press for the American Association for the Advancement of Science, 1979), p. 42.
2. Dorothy Nelkin and Judith P. Swazey, "Science and Social Control, Controversies over Research on Violence," in Hans Skoie, ed., *Scientific Expertise and the Public,* Conference Proceedings (Oslo, Norway: Institute for Studies in Research and Higher Education, The Norwegian Research Council for Science and the Humanities, 1979), p. 208.
3. Gerald Holton, "Epilogue to the Issue, 'Limits of Scientific Inquiry,' " *Daedalus,* vol. 107, no. 2 (Spring 1978), p. 227.
4. Keith M. Wulff, "Research Regulation, the Public, and Professional Organizations," in Wulff, p. 220.
5. *Ibid.*
6. Andre E. Hellegers, "The Ethical Dilemmas of Medical Research," in Wulff, pp. 9–10.
7. Panel discussion in *ibid.*, pp. 21–24.
8. Barry M. Casper, "Value Conflicts in Restricting Scientific Inquiry," *ibid.*, pp. 15–20.
9. Discussion of whistle-blowing "as a new American political phenomenon" that emerged in the 1960s is in Charles Peters and Taylor Branch, eds., *Blowing the Whistle* (New York: Praeger Publishers, 1972), chap. 1.
10. *Scientific Freedom and Responsibility,* A Report of the AAAS Committee on Scientific Freedom and Responsibility, prepared for the Committee by John T. Edsall (Washington, D.C.: American Association for the Advancement of Science, 1975), pp. 31–40.
11. Edward Weisband and Thomas M. Franck, *Resignation in Protest* (New York: Grossman Publishers, 1975), pp. 1–4.
12. Peters and Branch, pp. 280–81.
13. Joel Primack and Frank von Hippel, *Advice and Dissent* (New York: Basic Books, Inc., 1974), p. 253; Weisband and Franck, p. 183.
14. For an argument in favor of anonymous whistle-blowing, see Frederick A. Ellison, "Anonymous Whistle-blowing: An Ethical Analysis," *Business and Professional Ethics Journal,* vol. 1, no. 2 (Winter 1982), pp. 39–58.
15. Task Force of the Presidential Advisory Group on Anticipated Advances in Science and Technology, "The Science Court Experiment: An Interim Report," *Science,* vol. 193, no. 4254 (August 20, 1976), pp. 653–56; Arthur Kantrowitz, "The Science Court Experiment: Criticisms and Responses," *The Bulletin of the Atomic Scientists,* April 1977, pp. 44–50. A cogent critique of

the Science Court proposal, and alternative suggestions that parallel some made here, are in Barry M. Casper, "Technology Policy and Democracy: Is the Proposed Science Court What We Need?" *Science,* vol. 194, no. 4260 (October 1, 1976), pp. 29–35.

16. Task Force of the Presidential Advisory Group, p. 654.

17. Some supporters of the Science Court idea do not agree that the Court should decide on matters such as this, yet the Task Force report on "The Science Court Experiment . . ." includes this passage: "[W]e propose to prevent selection of a part of the issue which might prejudice the result. For example, the issue would not be, Are nuclear power plants explosive in the sense of an atomic bomb? but, Should a specific nuclear plant be licensed or not be licensed?" *Science,* vol. 193, no. 4254 (August 20, 1976), p. 654.

PART FOUR
Conclusion

10

Turbulence, Trends, and the Future

The 1970s were years of turbulence for science, and the 1980s promise to be no less so. Long-standing notions about science have come under challenge, and the earlier confidence that scientific discoveries would redound to man's benefit has yielded to skepticism. The promise of nuclear energy has been marred by the threat of nuclear accidents and the danger of radioactive waste. The effectiveness of antibiotics has diminished as resistant strains of bacteria evolve. The use of pesticides and food preservatives, intended to ameliorate the human condition, has revealed unexpected effects that threaten health. Exploring the causes of disease through recombinant DNA research may pose dangers resulting from genetic tampering. The anticipation a generation ago that science would soon herald a glorious age for the earth's inhabitants has receded. "The age of science triumphant," writes Jean-Jacques Salomon, "is also the age of absolute menace, of nonsense, of derision."[1] While most of the public apparently continue to believe that the benefits of science outweigh its harmful effects, the margin of confidence has fallen.[2]

At the same time that practical applications have come into question, so has the nature of science. Scientific method has been described by some as merely the application of common sense, by others as anything but common sense.[3] Until the early sixties the belief in scientific truth and universal laws was hardly questioned, but when Thomas Kuhn wrote in *The*

Structure of Scientific Revolutions that science is structured by paradigms, he started his own revolution.

A paradigm, according to Kuhn, provides the model problems and solutions at a particular time for a community of practitioners.[4] Pre-Copernican astronomy comprised a different paradigm from post-Copernican astronomy, Newtonian mechanics from Einsteinian, and pre-Darwinian assumptions from post-Darwinian. Problems to be investigated and discoveries to be made would therefore fall within the context of specific systems of understanding. If, as Kuhn professes, "there can be no scientifically or empirically neutral system of language or concepts,"[5] what happens to universal truth or law?

Where Kuhn is suggestive, Paul Feyerabend is decisive. He dismisses the idea that "pure unadulterated science gives a true account of man and the world." He "opposes positively and absolutely . . . universal standards, universal laws, universal ideas such as 'Truth,' 'Reason,' 'Justice,' 'Love.' "[6] Scientific truth is determined by vote, according to Feyerabend, by the decisions of the scientific establishment and by politicians, based on money, interest, and pride. The belief that facts, logic, and methodology reveal truth "is what the fairy-tale tells us."[7]

One thus senses the contemporary tumult in describing science. Science is and is not common sense. It is and is not based on universal truths and laws. It is and is not value-free, affected or unaffected by politics, culture, and myth. Every position draws supporters and critics.

Changing Attitudes within the Scientific Community

As the nature of science has been debated so also, as we have discussed throughout this book, has the appropriateness of external interference with scientific activity. Scientists who treat all political restraint as anathema, cling to an outmoded notion. They fail to understand critical differences between politically imposed truth, and unethical or hazardous activity. Our survey suggests that there is movement away from this position. In an evaluation of three age groups—39 and under, 40 through 59, 60 and over—scientists in the oldest group appeared most committed to traditional assumptions and beliefs, scientists in the youngest group the least.

Table 10.1 Scientists' Attitudes about General Scientific Controversies (Questions 7, 12, and 13)

	Percent Agree*		
	Youngest (age 39 and under)	Middle (age 40–59)	Oldest (age 60 and over)
A. Genuine scientific truth exists apart from human values.	86	93	94
B. Science more than other disciplines is structured on objective truths.	90	91	92
C. The attributes of a good scientist are universal and are independent of national or cultural influences.	77	82	83

*Includes respondents who indicated "strongly agree" or "agree." The other choices were "disagree," "strongly disagree," and "can't say."

ON THE NATURE OF SCIENCE

A large majority of the scientists questioned agree that "genuine scientific truth exists apart from human values," but the firmness of conviction appears to be proportionate to the respondent's age (Table 10.1:A). Although the pattern of overall agreement (strongly agree plus agree) is linear, it is accentuated among the 54 percent of the youngest group who strongly agree with the proposition, compared to 56 percent of the middle group, and 70 percent of the oldest.

The differences among age groups are narrowed in response to the statement that "science more than other disciplines is structured on objective truths" (Table 10.1:B). Nevertheless, the linear configuration by age is again emphasized by those who strongly agree: 24 percent of the youngest group, 27 percent of the middle, 35 percent of the oldest.

The pattern holds in the scientists' responses to the assertion that "the attributes of a good scientist are universal and are independent of national or cultural influences." Seventy-seven percent of the youngest group are in overall agreement with the proposition, 82 percent of the middle, and 83 percent of the oldest (Table 10.1:C).

While the differences among the age groups are not overwhelming, they are consistent. Most scientists adhere to the traditional belief that science is structured on objective truths apart from human values. But new considerations about the essence of science, introduced in the 1960s and intensely debated during the 1970s, appear to have influenced the thinking of younger scientists more than older.

ON SCIENTIFIC AND TECHNOLOGICAL ACTIVITIES

For assessments of dangers posed by specific scientific or technological activities, attitude differences based on a respondent's age become more pronounced. The consequences of four activities widely discussed during the seventies—recombinant DNA research, nuclear power, nuclear waste proliferation and disposal, environmental pollution—were consistently perceived as more hazardous by younger than older scientists.

During debates on these issues, specialists often tended to minimize the likelihood of hazard in their own fields. Most molecular biologists, for instance, concluded that recombinant DNA research was not dangerous; nuclear scientists by and large remained confident that nuclear power could be developed safely; and concerns about environmental pollution were frequently seen as exaggerated by environmental researchers.[8]

Without exception, as Table 10.2 shows, the oldest scientists accepted these assurances of safety in greater numbers than did the youngest. In assessing the risks to public safety of the four activities listed in the table, the responses approximated a linear pattern for the three age groups. The youngest group generally viewed each activity as more risky than did the middle group, and the middle group more than the oldest. The only exceptions appeared on the issues of recombinant DNA research and nuclear power plants. Fifty-one percent of both the middle and youngest groups termed recombinant DNA research very risky or moderately risky, while more of

Table 10.2 Scientists' Attitudes about the Risks of Four Activities (Question 19)

Certain scientific or technological activity may create risks to public safety. How do you feel about the following:

	Percent Risky*		
	Youngest (age 39 and under)	Middle (age 40–59)	Oldest (age 60 and over)
A. Recombinant DNA research (altering genetic code of a cell).	51	51	37
B. Existing levels of environmental pollution.	64	60	54
C. Nuclear waste proliferation.	81	79	64
D. Nuclear power plants.	40	46	29

*Includes respondents who indicated "very risky" or "moderately risky." The other choices were "slightly risky," "not at all risky," and "can't say."

the middle group (46 percent) than the youngest (40 percent) viewed nuclear power plants as very or moderately risky.

Otherwise, the linear relationships by age hold for perceptions about the hazards of specific scientific or technological activities, as they do for attitudes about the nature of science. The oldest scientists cling to traditional assumptions and to the words of experts, while the youngest appear more skeptical.

ON GOVERNMENTAL INTERFERENCE

The linear correlation of age to traditional beliefs is even more pronounced on the question of governmental interference with scientific activity. Only 30 percent of the youngest group of scientists agree (10 percent strongly) that "the government should never interfere with scientific activity that genuinely seeks to increase knowledge." Seventy-two percent of the oldest are in agreement (37 percent strongly), while 47 percent of the middle group are (15 percent strongly), as Table 10.3:A shows.

Table 10.3 Scientists' Attitudes about Science Policy Controversies (Questions 16, 20, and 22)

	Percent Agree*		
	Youngest (age 39 and under)	Middle (age 40–59)	Oldest (age 60 and over)
A. The government should never interfere with scientific activity that genuinely seeks to increase knowledge.	30	47	72
B. Restrictions imposed by the government on recombinant DNA research constitute unjustified interference with discovery of scientific truth.	20	27	51
C. If the work of nuclear scientists and technicians had not been impeded by politics external to their discipline, nuclear power could have been expanded more effectively and safely.	35	37	62

*Includes respondents who indicated "strongly agree" and "agree." The other choices were "disagree," "strongly disagree," and "can't say."

The reluctance on the part of the oldest group to see the government regulate, limit, or prohibit "genuine" scientific activity is consistent with the traditional view. The young scientists' apparent willingness to acknowledge that governmental interference is sometimes necessary, represents a departure from long-held values in the community of science. The age-related pattern is apparent in responses to questions about governmental interference with recombinant DNA research and nuclear power development (Table 10.3:B and C).

GENERATIONAL EXPERIENCE: AN EXPLANATORY MODEL

The consistent responses of scientists according to age are probably related to behavioral phenomena arising either from changes in attitudes of individuals during their lives or from important generational experiences.[9] The weaker of the two suppositions seems to be that younger scientists at any moment are more skeptical than their elders and that they will become more tradition-oriented as they grow older. According to this view, the oldest scientists in our study would have responded more skeptically when they were younger, but their attitudes changed with increasing age. Analogies in political behavior support the inference of age-related change insofar as there is a modest, though frequently overstated, tendency for people to become more conservative as they grow older.[10]

A study by Ladd and Lipset and another by Anand and Haberer correlated the political attitudes of scientists with several variables. Although their findings differed in some respects—Ladd and Lipset, for example, found physicists to be the most liberal among academic scientists, while Anand and Haberer did not—both reported that younger scientists tended to be more liberal than older.[11] Neither study sought to explore the basis of their age-related findings.

In any case, one may argue that political ideology bears little relationship to the questions about science raised in this study.[12] Moreover, psychological and political studies support generational experiences as a likely explanation for the age correlations. Many people, whether they were young adults during the Depression, World War II, or the Vietnam War, developed political and social attitudes as a result of their experiences that became fixed for the rest of their lives.[13]

Of course the reasons for the age-related correlations in this and other studies can be judged conclusively only after surveying the same scientists through several decades. Short of such an ambitious undertaking, we find inferential evidence that generational experiences have been the principal determinants of skepticism among the younger scientists in our survey. This is based on recognition that the views of the older scientists parallel those of the prevailing ethos of an earlier generation; most of the older group apparently believe now as they did then.

Anthony Standen wrote in 1956 that the public regarded scientists as "lofty and impeccable. . . . Since it is only natural to accept such flattery, the scientists accept the laymen's opinion about themselves."[14] The scientist whose formative professional years occurred during World War II and the decade following is unlikely now to shed the image of grandeur about science and scientists imbedded earlier. On the other hand, those who are now in their twenties and thirties were exposed to far more skepticism about science. As a result, they may have developed different attitudes from their predecessors.

Observers less committed to the earlier ethos, including the younger scientists, may now more easily sense its implausibility. They are likely to notice the contradiction that scientific truth was supposedly unrelated to human values, but that discovering these truths was inherently good. "If you are a scientist you believe that it is good to find out how the world works; that it is good to find out what the realities are," said Robert Oppenheimer in 1945.[15] "Good" is a value judgment, and if scientific understanding is good, then understanding science has been inexorably tied to human values.

This developing recognition may also be linked to views about the conduct of scientific activity and to the appropriate role of political authority. The argument is consistent with Martin Brown's observation that "it is the young scientist who is most aware of the failure of science and most willing to do something about it."[16] The more one recognizes the human relationship to scientific activity, the greater the sense of responsibility for its consequences, harmful as well as beneficial. Increased receptivity to governmental intervention should therefore not be surprising because the government is the ultimate instrument devised by society to protect itself from danger.

IMPLICATIONS FOR THE FUTURE

If the pattern of responses by scientists according to age is largely based on generational experiences, as suggested here, several consequences may be anticipated. As the public's earlier view of the grandeur of scientists helped shape the attitudes of scientists, so will continuing public skepticism also influence the self-image of future scientists. An increase in the proportion of scientists who reject the traditional assumptions and values may therefore be expected. In attitudes about the three categories—the nature of science, specific activities, governmental interference—the number of scientists whose positions parallel those of the youngest scientists in this study will continue to grow.

If more scientists relate science and its understanding to human values, scientific hazards will be seen increasingly as human responsibilities. In consequence, as with younger scientists in this sample, the scientific community may become less antipathetic to governmental regulation. The trend in each of the categories dealt with in the survey appears to be moving in that direction. In addition, as more scientists assimilate the new values, they will reinforce public consciousness of the hazards as well as the benefits of science and technology, thus increasing public sensitivity and skepticism as experts confess their own uncertainties.

Before an equilibrium is reached, continuing ferment and skepticism among the public and within the scientific community is to be expected. Any equilibrium will likely rest on the notion that scientific activity should be subject to constraints no greater and no less than for other activities that affect the public welfare.

The Proper Role of Science—A Competing Interest

Neither abnegation nor overbearance on the part of government is desirable. George Ball's dictum that scientists should "decide what research to undertake and how to undertake it, subject only to such safeguards as they might individually or collectively impose,"[17] can hardly be justified in view of experiences cited in this book. Nor is there comfort in Peter Hutt's stifling admonition that "the role of scientists should be limited to conducting the type of research that the public concludes to be relevant to its needs (like the role of generals in conducting wars)."[18]

Rather we should understand that scientists comprise one among many interest groups. Many observers shrink from this recognition. Some have referred to science as a sacred establishment and to scientists as a priesthood or ecclesiastical order.[19] Neither science nor scientists (nor anyone else) deserve such an exalted position in a democracy. The scientific community is an interest group that should have to compete in the political and economic marketplace as do lawyers, teachers, and businessmen. Despite entreaties by some, scientists should no sooner be the exclusive regulators of science than farmers should be of agriculture or company executives of business.

Though this idea is unsettling to those who seek neat, circumscribed answers, dealing with science in a democratic society cannot be tidy. Pluralist politics is unavoidably imprecise. As Alexander Rich recognized at a symposium on managing science, if the problems are handled "in a pluralist manner, with many different groups having control of particular sectors in which they are interested, we will in fact muddle through in a way that is very much consonant with our democratic principles."[20]

But "muddling through" does not preclude the need for occasional innovation. Whether through a Science Hearings Panel as proposed in this book, or in some similar manner, a systematic means to expose potentially disastrous scientific activity seems essential. As scientific and technological activities continue to multiply, this need becomes more compelling. The structures and values that underlie the American political system are suited to the purposes of such a panel. Openness, an educated polity, and a commitment to the people's welfare are central to the American political culture. To the extent that a panel would foster these attributes, the citizenry would be better protected from a variety of excesses related to science—from externally imposed truth, from unethical research, from masking the dangers of certain activities.

The threat of irreversible hazards may best be dealt with when understood as another in a long list of threats to liberty. Protection from them will require no less vigilance than has been required against abuse of individual liberties since the founding of the nation. No miracle institutions can insure protection from such awesome threats. Contrary to what Willard Libby said when he was a member of the Atomic Energy Commision, the American (and the world's) population should not have to learn to live with radioactive fallout. We do, however, have to live with its threat.

The thought is sobering but not hopeless. Neither science nor scientists will save us from externally imposed scientific truth or from irreversible hazards. Only the nexus of structures and values that comprises the political system can. Not regulatory agencies, democratic elections, enlightened citizens, a free press, a Science Hearings Panel, or any other single institution can insure protection. They are all essential.

Predicting the future of the relationship between science and politics is particularly risky amid the ferment that attends the issue today. Our analysis encourages confidence that the American polity is immune to perversions on the order of the historic cases. Unfortunately the frequency with which these incidents are linked by scientists and scholars to events in America blurs the issue that should be addressed: whether and how the political system can protect us against science and technology that threaten society with catastrophe.

The need for governmental interference, increasingly acknowledged within the scientific community, demonstrates a willingness by experts to question outworn assumptions. While many scientists cling to the traditional notion that science and scientists should remain unrestrained, others recognize changed conditions. The threat of irreversible hazards, the potential dangers in some activities, and public concern have all served to heighten the sensitivity of many scientists to issues that they had previously ignored. We may hope that this emerging awareness among scientists and the public will spur the national effort to confront the problems.

Notes

1. Jean-Jacques Salomon, *Science and Politics* (Cambridge, Mass.: The M.I.T. Press, 1973), p. 241.

2. L. John Martin, "Science and the Successful Society," *Public Opinion* (June/July 1981), p. 18. Public confidence in institutions in general has fallen since the 1960s, and confidence in science remains higher than in most other institutions. See Allan Mazur, "Commentary: Opinion Poll Measurement of American Confidence in Science," *Science, Technology, and Human Values,* vol. 6, no. 36 (Summer 1981), pp. 16–19; and Kenneth Prewitt, The Public and Science Policy," *ibid.*, vol. 7, no. 39 (Spring 1982), pp. 6–7.

3. Scientific method, according to Bernard Dixon, "amounts to no more than scrupulously applied common sense." *What Is Science For?* (New York: Harper and Row, 1973), pp. 34–35. Similarly see J. Bronowski, *The Common Sense of Science* (New York: Vintage Books, n.d.); and James B. Conant, *Science and Common Sense* (New Haven, Conn.: Yale University Press, 1974). Conversely, Hanna Pitkin writes that scientific knowledge has "left common sense and ordinary human understanding hopelessly behind." *Wittgenstein and Justice* (Berkeley, Cal.: University of California Press, 1974), p. 318.

4. Thomas S. Kuhn, *The Structure of Scientific Revolutions,* 2d ed. (Chicago: University of Chicago Press, 1970), p. viii.

5. *Ibid.,* p. 146.

6. Paul Feyerabend, *Against Method* (London: NLB, 1976), pp. 188–89,

7. *Ibid.,* pp. 302–3.

8. Responses by scientists to the questionnaire demonstrate the tendency of those closest professionally to issues of controversy to appear more confident about safety. Thus while 59 percent of the biologists believe that recombinant DNA research is slightly or not at all risky, only 46 percent of the physicists do. Conversely, 61 percent of the physicists consider nuclear power plants to be slightly or not at all risky, but only 53 percent of the biologists do. This parallels Allan Mazur's findings that biomedical scientists more than physicists tend to oppose nuclear power. He ascribes this phenomenon to the influence of social networks within disciplines rather than to expertise. *The Dynamics of Technical Controversy* (Washington, D.C.: Communications Press, 1981), pp. 80–82.

9. In addition to age, crosstabulations were made for discipline (physicist, biologist, other), employment (academic, industry, government), and research activity (involved/not involved). Only age, however, yielded consistent patterns relative to the issues described in this study.

10. Robert E. Lane, *Political Life* (Glencoe, Ill.: The Free Press, 1959), pp. 217–18; Angus Campbell, Philip E. Converse, Warren E. Miller, and Donald E. Stokes, *The American Voter* (New York: John Wiley and Sons, 1960), pp. 210–11; Norman H. Nie, Sidney Verba, and John Petrocik, *The Changing American Voter* (Cambridge, Mass.: Harvard University Press, 1976), pp. 263–67. The studies caution against exaggerating the tendency, however.

11. Everett Carll Ladd, Jr. and Seymour Martin Lipset, "Politics of Academic Natural Scientists and Engineers," *Science,* vol. 176 (June 9, 1972), p. 1097; Hans Raj Anand and Joseph Haberer, "Scientific and Political Orientation of American Scientists," *Research Policy,* vol. 7 (1978), pp. 29, 41.

12. Though opponents of some scientific or technological activities may tend to be conservative (as with fluoridation), or liberal (as with nuclear power), or mixed (as with high voltage transmission lines), there is nothing "intrinsically liberal or conservative" about such activities, as Allan Mazur observes. Mazur, pp. 46–47, 60–61.

13. Campbell, et al., pp. 151–55; Nie, et al., pp. 74–95. The concept of generational experience is further elaborated on in Erik H. Erikson, *Young Man Luther* (New York: W. W. Norton and Co., 1962), p. 41; and Seymour Martin Lipset, *Political Man* (Garden City, N.Y.: Doubleday and Co., 1960), p. 265.

14. Anthony Standen, *Science Is a Sacred Cow* (New York: E. P. Dutton and Co., 1956), pp. 14–15.

15. Alice Kimball Smith and Charles Weiner, *Robert Oppenheimer, Letters and Recollections* (Cambridge, Mass.: Harvard University Press, 1980), p. 317. More recently Philip Handler, while president of the National Academy of Sciences, chided all who doubt the assumption that "objective knowledge is an unquestioned good." *Science,* vol. 208, no. 4448 (June 6, 1980), p. 1093.

16. Martin Brown in *The Social Responsibility of the Scientist,* Martin Brown, ed. (New York: The Free Press, 1971), p. 271.

17. Quoted in Sissela Bok, "Freedom and Risk," *Daedalus,* vol. 107, no. 2 (Spring 1978), p. 118.

18. Peter Barton Hutt, "Public Criticism of Health Science Policy," in *ibid.,* p. 163.

19. Standen, *Science Is a Sacred Cow;* Don K. Price, *The Scientific Estate* (New York: Oxford University Press, 1965); Ralph E. Lapp, *The New Priesthood* (New York: Harper and Row, 1965). The authors, however, generally expressed wariness about popular acceptance of science in such lofty terms.

20. In "Commentaries from a Series of Academy Forums," *Science: An American Bicentennial View* (Washington, D.C.: National Academy of Sciences, 1977), p. 75.

Appendix
The Survey and Interviews

In October and November 1979, questionnaires were mailed to 1,420 scientists, randomly selected from the 1976 edition of *American Men and Women of Science.* Eighty-two letters were undeliverable and were returned unopened. Seven questionnaires were inadequately filled out, while 632 were satisfactorily completed and returned. The response rate from those who received questionnaires was about 48 percent.

The overall sample of 1,420 included 16 percent who were age 39 or under, 63 percent between 40 and 59, and 21 percent over 60. Among the 632 who returned usable responses the pattern was similar: 18 percent in the youngest group, 63 percent in the middle, and 19 percent in the oldest (percentages were rounded off). Division by disciplines was comparable, though physicists were underrepresented among the oldest respondents. Among biologists (N=184) 23 percent fell in the youngest category, 19 percent in the oldest; among chemists (N=170) 15 percent and 18 percent; among physicists (N=80) 20 percent and 9 percent.

Questionnaire items largely involved short answers or multiple choices. Response rates for substantive answers can be determined by subtracting the "no answer/can't say" figure in each table from the total number of respondents in the table. Respondents also were invited to add comments. Many did, including some who appended several pages to amplify their views. Samples of these comments have been incorporated into the text where appropriate.

In addition, I interviewed more than one hundred scientists and others who have been involved with developing science

Table A.1 Selective Characteristics of the Survey Sample (632)

Age	Percent
Under 30	<1
30–39	18
40–49	30
50–59	32
60–69	16
70 and over	3
Field	
Biology	29
Medicine/health	12
Physics	13
Chemistry	27
Other physical sciences (geology, anthropology, mathematics, etc.)	15
General/ecology/nonscientific	4
Highest degree	
PhD/DrSci/DEd	83
MD/DDS/DVM	10
MA/MS	5
BA/BS	2
Primary Involvement	
Research	27
Administration	11
Teaching	5
Research and teaching	45
Research and administration	7
Private practice/consultant/retired	3
No answer	2
Principal employer	
College/university	63
Private industry	21
Government	9
Self	4
Other	3

policies in the United States. Many had reputations for expertise about issues discussed in the book. In areas of controversy, such as nuclear or recombinant DNA policies, I sought out people who represented a variety of viewpoints. Several interviews included solicitation of short-answer responses as in the questionnaire survey, while others were less structured. Most ranged between thirty minutes and an hour, though some lasted several hours. The interviews provided opportunities to explore more fully the questions and issues that were treated briefly in the questionnaire.

QUESTIONNAIRE

351A, Science Building
Project on Politics and Science
William Paterson College of New Jersey
Wayne, New Jersey 07470

PERSONAL BACKGROUND

1. Year of birth ____________________

2. Field of specialization ____

3. Highest degree(s) obtained ____
degree(s) school(s) year

4. Present position primarily involves *(please check)*
1. ☐ administration 2. ☐ research 3. ☐ teaching-research 4. ☐ other

5. Principal employer is
1. ☐ college/university 2. ☐ private industry 3. ☐ government 4. ☐ self 5. ☐ other ________

6. Consider the scientific and technical activities for which you are responsible (your own work, or work you supervise). Who has influence in deciding the nature or objectives of your work?
Estimate the percent of influence by each of the following to the nearest 5-10%.

	Percent influence in deciding nature or objectives of work
Myself	________ %
Colleagues (without supervisory authority over me)	________ %
My immediate chief	________ %
Higher level supervisors or authorities	________ %
Other ______________________________ *(please fill in)*	________ %

Total should add to 100%

Please check the box following each statement that best describes your reaction, even if based largely on impressions. Comments are welcome in the margins or on additional paper.

		Strongly Agree	Agree	Disagree	Strongly Disagree	Can't Say
		1	2	3	4	5
7.	Genuine scientific truth exists apart from human values. . .	☐	☐	☐	☐	☐
8.	The imposition of scientific truth by a political system and the prohibition of challenges to that truth may aptly be characterized as a perversion of science.	☐	☐	☐	☐	☐
9.	A scientist's training and experience provide an advantage over nonscientists in ability to analyze issues. . . .	☐	☐	☐	☐	☐
10.	Scientists are more likely than nonscientists to reject unsubstantiated claims of truth or fact.	☐	☐	☐	☐	☐
11.	Scientists work most creatively when the nature and objectives of their research are defined in part by higher authorities, and not when they are completely self-determining.	☐	☐	☐	☐	☐
12.	Science more than other disciplines is structured on objective truths.	☐	☐	☐	☐	☐
13.	The attributes of a good scientist are universal and are independent of national or cultural influences.	☐	☐	☐	☐	☐
14.	Historic perversions of science like Nazi racial science, Soviet biology under Lysenko, and unchallengeable dogma that the earth was the center of the universe would be virtually impossible to impose in American society.	☐	☐	☐	☐	☐

15. The most important safeguard in American society against the imposition of a scientific perversion is
1. ☐ the nature of the American political system 2. ☐ the values and beliefs of Americans in general
3. ☐ the nature of most American scientists 4. ☐ the way in which science is organized
5. ☐ other ________________ 6. ☐ there are no safeguards 7. ☐ can't say

	Strongly Agree 1	Agree 2	Disagree 3	Strongly Disagree 4	Can't Say 5
16. The government should never interfere with scientific activity that genuinely seeks to increase knowledge.	☐	☐	☐	☐	☐
17. The prohibition of teaching the theory of evolution in several states constituted a scientific perversion comparable to the historic cases cited in question 14.	☐	☐	☐	☐	☐
18. The use of involuntary human subjects for medical experimentation constitutes a perversion of science.	☐	☐	☐	☐	☐

19. Certain scientific or technological activity may create risks to public safety. How do you feel about the following *(please check).*

	Very Risky 1	Moderately Risky 2	Slightly Risky 3	Not At All Risky 4	Can't Say 5
Recombinant DNA research (altering genetic code of a cell). .	☐	☐	☐	☐	☐
Nuclear waste proliferation	☐	☐	☐	☐	☐
Nuclear power plants	☐	☐	☐	☐	☐
Existing levels of environmental pollution	☐	☐	☐	☐	☐
Other ______________________ . . *(please fill in)*	☐	☐	☐	☐	☐

	Strongly Agree 1	Agree 2	Disagree 3	Strongly Disagree 4	Can't Say 5
20. Restrictions imposed by the government on recombinant DNA research constitute unjustified interference with discovery of scientific truth.	☐	☐	☐	☐	☐
21. As I understand it, creating a safe permanent repository for nuclear waste is not a major technological problem. . . .	☐	☐	☐	☐	☐
22. If the work of nuclear scientists and technicians had not been impeded by politics external to their discipline, nuclear power could have been expanded more effectively and safely.	☐	☐	☐	☐	☐

23. What should a scientist do when he fears that the consequences of scientific activity with which he is familiar might be disastrous? Should he

1. ☐ engage in other activity and say nothing 2. ☐ raise concerns in public
3. ☐ raise concerns only within the scientific community 4. ☐ other ______________ 5. ☐ can't say

24. What do you think most American scientists would do under the circumstances proposed in the previous question? Would they

1. ☐ engage in other activity and say nothing 2. ☐ raise concerns in public
3. ☐ raise concerns only within the scientific community 4. ☐ other ______________ 5. ☐ can't say

25. The principal determinant of whether a scientific perversion might be imposed in any society lies in the nature of

1. ☐ the scientific discipline in question 2. ☐ scientists in that society
3. ☐ the organization of scientists in that society 4. ☐ the political system 5. ☐ other ______________
6. ☐ can't say

26. Scientific activity that may be hazardous to the public should be

1. ☐ regulated by the government 2. ☐ regulated by designated scientific experts
3. ☐ not formally regulated beyond the good sense of the experimenter 4. ☐ other ______________
5. ☐ can't say

27. How does one decide when governmental interference in scientific activity is legitimate or not? *(please comment)*

Index